sey

r

ks

The Clipper Wildfire *about to sail from Birkenhead to The Crimea, 21 December 1854.*

'The roads you travel so briskly lead out of dim antiquity, and you study the past chiefly because of its bearing on the living present and its promise for the future.'

Lieutenant General James G. Harboard,
K.C.M.G., D.S.M., LL.D., U.S. Army (Retd.)
Late American Member of Council in London
The Newcomen Society of England

'History is bunk.'
Henry Ford

ISBN 0 9510965 2 4 Thomas Brassey

British Library Cataloguing-in-Publication Data
A catalogue record of this book is available
from the British Library

Extracted from "William Heap and His Company"

First Published 15th October 1976
Second Edition April 1977
Third Edition October 1991
This Extract First Published April 1993

Published by John Millar (UK) Ltd, Hoylake,
Wirral, Cheshire, England

Printed by G.M. Business Print, Liverpool

Set in Monophoto 400 Century

INTRODUCTION

This booklet is an extract from the Company history, "William Heap & His Company" and describes the early life of William Heap, who was Thomas Brassey's Manager of the Bridge Department at Canada Works, Birkenhead, from 1833 until 1866 and who later became a Director of Canada Works Engineering & Shipbuilding Company.

Thomas Brassey (1805–1870) was the greatest Railway Contractor in the World and in the 1850's he employed no less than 85,000 men, all superb workmen, on contracts all over the World.

In 1853 he set up CANADA WORKS, on the Great Float in Birkenhead initially to build the locomotives, wagons and bridges for the Grand Trunk Railway of CANADA, but later for many other contracts as well.

William Heap worked for Thomas Brassey from the time CANADA WORKS was built in 1853 until he left to set up his own company in 1866. Later he became a Director of CANADA WORKS ENGINEERING LTD.

The feats carried out by Thomas Brassey and his men were breathtaking. They built railways all over the United Kingdom and on every Continent. They helped to open up India, Canada, Australia and South America and so changed social patterns, for the better, all over the World.

In 1850 they built the longest bridge in the World. In 1855 they saved the starving, disease ridden, badly led British Army of 30,000 men before Sebastopol in the Crimea by building a 39 mile long railway from Balaclava to all parts of the front before Sebastopol in the incredible time of 60 days from the order being given.

Thomas Brassey and his men working out of CANADA WORKS, Birkenhead, were Victorian Giants and we all owe them a great debt of gratitude because more than 100 years later we, and millions of others all over the World, still benefit from their tremendous achievements.

John Millar
Heap & Partners Ltd, Engineers,
Hoylake, Wirral.

CONTENTS

ACKNOWLEDGEMENTS

I have received much help in the preparation of this book and a full acknowledgements list appears in the full book "William Heap & His Company".

Part of the cobbled road that ran through Canada Works, Birkenhead, relaid in front of the offices of Heap & Partners Ltd, Hoylake, Wirral in 1971.
The plaque tells passers by something of Thomas Brassey and his men working out of Canada Works.

CHAPTER I

Early Days

HEAP FAMILY TREE

William Heap (Grocer) (b. 1760 d. 1822) married (1781) Nancy . . . (b. 1756 d. 1822)

- John (b. 1783 d. 1819)
- William (Weaver) (b. 1784 d.)
- George (b. d.)
- Mary (b. d.)
- James (b. 1795 d. 1823)

William (Weaver) Married (29-8-1822) Mary Smith of Burnley

- Elizabeth Mary (b. 1825)
- William (b. 1826 d. 1912)
- Edmund (b. 1827 d.)
- John (b. 1831 d.)
- James (b. 1834 d.)
- Ann (b. 1836 d.)

William Married (1-6-1859) Eliza Mackenzie Burn Stuart Crichton (b. 1835 d. 1907)

- William Crichton (b. 1861 d. 1871)
- Charles Edmund (b. 1865 d. 1881)
- Walter Marillier (b. 1875 d. 1876)
- Mary Winifred (b. d. 1912)
- Elizabeth Ann (b. d. 1939)

PADIHAM 1760–1800, RIPE FOR REVOLUTION

In the north of Lancashire lies the small town of Padiham, built on a hill rising from the banks of the River Calder with Pendle Hill to the north, home of witches and warlocks. The town is industrial but has a certain charm because it is on the edge of some beautiful country and the older houses are all built of the grey local stone, weathered, strong and enduring like the people of Lancashire.

In 1760, when the first William Heap was born, Padiham was a little village in a lovely setting with rolling hills on all sides. The quiet rhythm of life of its inhabitants had not changed for centuries but elsewhere in England men were working on new inventions that were to change the life of the Heap family, Padiham and eventually the whole world. The old pastoral way of life was about to be changed by a revolution.

When the first William Heap was seven years old not many miles away James Hargreaves invented the Spinning Jenny; the year after Arkwright made his water frame in Preston, the following year in 1769 James Watt patented the separate condenser for steam engines, which vastly improved the clumsy, inefficient engine first built by Thomas Newcomen in the early part of the century. When the first William was fifteen, two events occurred which primed the bomb of the industrial revolution, in Bolton, Samuel Compton invented the spinning mule and James Watt teamed up with Matthew Boulton and they started to produce more efficient engines at their Soho Manufacturings near Birmingham.

Padiham like other towns in Lancashire had coal for fuel, stone for building, the right climate for spinning cotton but, most important, had a hard working,enterprising people. It was ripe for the revolution which shook, and is still shaking the world, which enabled our small country to build the greatest empire the world has known, which opened up dark continents and changed society all over the world. It brought great benefits and great destruction and eventually enabled man to reach out to the stars and beyond.

WILLIAM HEAP (I) 1760–1822

Until the railways opened up the country, almost everybody with the surname Heap lived in Lancashire and the Public Records Office at Somerset House in London show that in 1832 they were not only almost all in Lancashire, but in general they nearly all lived within ten miles of Burnley.

Our founder's grandfather was a grocer and when he was twenty one he married a twenty-five year old girl called Nancy. They had five children, John, William (father of our founder), Mary, George and James.

The grocery business was a small family shop and it served the quiet, modest needs of the small country village, supplying flour, local cheese, black puddings, tripe, smoked ham and other local produce.

WILLIAM HEAP (II) 1784

It seems likely that the first William Heap's eldest son John was intended to carry on the family business because William initially decided to go into the new industry which was just starting up—cotton, and he became a weaver in one of the local mills.

On the 29th August 1822 William married Mary Smith of Burnley and they set up house in one of the stone built cottages in Padiham. Shortly afterwards, William's father died and as William's brother John had died three years earlier William left the mill and took over the family grocery business. William and Mary had six children between 1825 and 1836, Eliza, William (our founder), Edmund, John, James and Ann.

The coming of the cotton industry to Padiham and the growth of other industries, such as mining and quarrying in the area made a great impact on the village. The new industries attracted workers from the surrounding countryside and even as far away as Ireland. In 1800 the population of Padiham was 250 but by 1822 it had grown to 6,000 because of the new influx.

The rapid growth of population increased the demand for supplies of all kinds including groceries and the growth of the Heap business kept pace with the increasing demand. Within a matter of months William started to mill his own flour to keep up with the demand and then he started as a corn dealer as well so that he was able to buy corn from the farmers, mill it into flour and then retail. By the time our founder William Heap (III) was born the Heap family business had grown so much that his father was no longer a grocer but a prosperous merchant.

Isles House, Padiham. The childhood home of William Heap

The increase in the family fortune must have been as rapid as the progress of the Industrial Revolution itself because our William was born in Isles House, Padiham, a large, if not pretty, house and within ten years his father also owned half a dozen cottages in Padiham and rented an estate of 75 acres of lovely rolling parkland which was owned by the local Lord of the Manor, Le Gendre Pierce Starkie who lived in nearby Huntroyde Hall.

Isles House formed part of the Starkie estate Huntroyde Demesne and in the 1820's the Heap family lived in some style with their own coach, horses in the stables at the back of the house and several servants.

William Heap Senior must have been on friendly terms with Pierce Starkie because the 75 acres that Mr Heap rented swept down quite close to the front of the magnificent country mansion Huntroyde Hall, the home of the Starkie family. Although it may seem surprising that a weaver turned grocer, turned merchant in the short time of five years should be on good terms with an ancient family it should be remembered that for centuries Padiham had been quite isolated in the hills and everybody knew each other in such a small village and there was mutual respect between the classes in this country in spite of what the apostles of class hatred would have us believe.

HUNTROYDE DEMESNE

The Starkie family originated in Cheshire in Norman times and there were branches of the family in Stretton and Barnton. In 1464 Edmund Starkie of Cheshire had the good sense to marry Elizabeth de Symonstone and through her gained possession of considerable estates in the Padiham area at Simonstone and Huntroyde Demesne.

Huntroyde Hall is a gracious building set in a natural bowl, the surrounding parkland is like

The 75 acre estate that William Heap's father rented from Le Gendre Pierce Starkie who lived in Huntroyde Hall, seen in the background, with Pendle Hill beyond.

The entrance gates from the land at the back of Isles House leading to the Huntroyde Demesne land rented by the Heap family.

so many other English country estates, a little gem with well positioned groups of elm and other trees. The estate is still in excellent order and the Starkie family still live there, although they have had to reduce the size of the Hall and have had to sell off part of the estate presumably to meet death duties. (Sadly in 1982 the Starkie family had to sell Huntroyde Hall after 500 years continuous occupation by the same family.)

WILLIAM HEAP (III) 1826–1912

Such was the background of the founder of our company, a large house with prosperous middle class parents, servants, horses and 75 acres of beautiful parkland to roam in with his brothers and sisters and other children of the village. The industrial revolution might be under way outside their grounds but that was unimportant to a child when there was Huntroyde Brook to fish in and horses to ride like the wind over lovely parkland.

SCHOOL DAYS

When William was old enough it was decided that he should go way to school and he was sent as a boarder to the best school in the area, Whalley Grammar School, situated some four miles from his home in Padiham.

The school was founded in 1549 by Edward VI the boy King son of Henry VIII and Jane Seymour. Edward was born in Hampton Court and succeeded his father to the thone in 1547 when he was only nine years old and was crowned in Westminster Abbey by Archbishop Cranmer. He was a sickly boy and the country was ruled by his Guardians, but one of the lasting monuments to his memory is the enormous number of schools he founded. The best

Whalley Grammar School painted in 1840 by a twelve year old pupil, J. G. Booth, who attended the school at the same time as William Heap.

ROYAL GRAMMAR SCHOOL,
WHALLEY,
N.E. LANCASHIRE.

known is Christ's Hospital, the Bluecoat School in London, founded four years after Padiham Grammar School and where the boys still wear the same uniform designed in Edward's day.

Edward VI granted Letters Patent to Whalley Grammar School in the second year of his reign and like many other splendid schools founded by Edward it gave a sound education to pupils for over four centuries. The school is still standing but although it has a Board of Governors it ceased to function as a school in the early days of the first war, but there are still one or two elderly residents living in Whalley who attended the school before it stopped taking pupils and now the school is used as the local Community Centre and Adult Education Centre.

These schools are part of our heritage handed down to us over four centuries and how any community can allow these schools to be destroyed by petty minded politicians, as they are in some areas, is beyond comprehension.

When the day came for William to leave Isles House for Whalley, his trunk was put in the family carriage and the horse trotted through the leafy road to the ancient village of Whalley and there he was left in the care of Mr and Mrs George Preston, the schoolmaster and his wife. Judging by letters sent by Joshua Ashworth, a pupil at the school, to his parents in 1853, the Prestons seemed to be quite reasonable people, but rumour has it that not all the schoolmasters at Whalley Grammar School were so inclined.

Even though Mr and Mrs George Preston were not unkind, William felt lonely in his first few days at the school and he missed his parents, brothers and sisters, but they were worked quite hard and he soon settled down.

The school was small by today's standards and like most of our best and most expensive boarding schools today, the emphasis was more on education than creature comforts, whereas in State schools the opposite is true and the country suffers accordingly. The floors in the school were flagged, the heating was by open coal fires and lighting was by oil lamps and candles. In the yard there was a hand water-pump and a stone trough and it must have been

William Heap's classroom. The concentration was on education not comfort. The gas lights were installed after William Heap had left.

The woodwork room at Whalley Grammar School. Here William Heap first learned to handle tools.

invigorating, to say the least, to go out into the cobbled yard on an icy winter's morning to draw water for washing from the well.

There were two classes of boarder's, those who stayed for the whole term and weekly boarders and it is likely that William Heap returned home to Isles House each weekend. The fees were £32 a term for the weekly boarders under ten, and £38 a term for those over ten. The extra for those boarders for the whole term was £4 a term.

There was no nonsense about the curriculum and William studied Holy Scripture, Reading, Writing, Arithmetic, Geography, History, Grammar, Composition, Literature, Euclid, Algebra, Trigonometry, Mensuration, Land Surveying, Latin, French, German, Greek, Natural Science, Chemistry, Mechanics and Drawing (Freehand, Model, Mechanical and Shading), Elocution, Book-keeping, Drill and Vocal Music and there were a number of optional extras—Woodwork, Dancing, etc. The excellent education William received at Whalley Grammar School stood him in good stead in his later life.

Whalley is a pleasant village, and William enjoyed the walks in the surrounding countryside and over the ruins of the beautiful Whalley Abbey. On those Sundays when he did not go home, William went with the other boys in a "crocodile" to Whalley Parish Church to worship. There the boys occupied certain pews reserved for the pupils of the Grammar School and he was able to see the Starkie family in their special pew at the front of the congregation.

So William spent his schooldays, working hard, but enjoying life as well. Towards the end of his school days the question of his future career was raised by his father and as William was the eldest boy, it was hoped that he would go into the prosperous family grocery business, but William told his parents that he wanted to become an Engineer, a decision he never regretted. He would undoubtedly have made a larger fortune if he had entered the family business, but he would not have led such an interesting life.

The Royal Grammar School, Whalley, as it is now.

APPRENTICESHIP

After leaving school, William Heap served an engineering apprenticeship but at this time we do not know the name of the firm, and we may never know. His subsequent career leads one to think that it may have been at a works similar to William Fairbairn's famous Canal Street Works, Ancoats, Manchester, but there are no early records of Fairbairn's apprentice's available apart from the more famous ones such as Albert Escher son of the founder of the Zurich factory.

We do know that after serving his apprenticeship William Heap saw an advertisement by William Evans of Cambridge who had secured the contract to build the tubular bridge at Conway and that William Heap was offered a position as a junior engineer with Evans and moved to the lovely little Welsh town of Conway in 1846 when he was just twenty years old, and where he was to meet some of the great giants of the Industrial Revolution, Robert Stephenson, Robert's father George, Brunel, Fairbairn and others

CHAPTER 2

The Conway, Britannia and Drogheda Bridges

Three masterpieces side by side. Edward the First's Conway Castle, Telford's Suspension Bridge and Robert Stephenson's Tubular Bridge—the world's first great tubular bridge.

CONWAY

The small Welsh coastal town of Conway is remarkable in having three great masterpieces side by side, all perfect jewels in their own right. Edward the First's castle, Thomas Telford's suspension bridge and Robert Stephenson's railway bridge, the first great tubular bridge ever built.

CONWAY CASTLE

Conway Castle was one of the chain of castles built by Edward I in Wales, but it is uncomparably the most magnificent. It was designed by James of St George, Master of the King's Works in Wales who was the greatest military architect of the age and was built by the greatest builder in the country, Richard of Chester.

Conway Castle is one of the most outstanding achievements of medieval military architecture in all Europe and men came from all over England and Wales to build it. In 1285 the labour force was 1,500 men and the chief joiner was Robert of Frankby who came from the small village of Frankby in Wirral about two miles from our Hoylake office.

TELFORD'S SUSPENSION BRIDGE

When Thomas Telford was given the task of building the A5 to improve the links between London and Holyhead, and hence Ireland, he built the graceful suspension bridge over the Menai Straits, but the Government also commissioned him to build a similar bridge over the river Conway, because a few years earlier a mail boat had capsized in the river and several local people had drowned.

Telford chose to build suspension bridges and the beauty of the wrought iron masterpieces are a joy to behold to this day.

ROBERT STEPHENSON'S RAILWAY BRIDGE

Robert Stephenson was retained as the Engineer for the Chester to Holyhead Railway which was started in 1845 and the greatest engineering problems were crossing the River Conway and the Menai Straits between Caernarvonshire and the Isle of Anglesey.

After surveying the land, Stephenson decided to build his railway bridge close to Telford's bridge at Conway, which meant that it would have to pass very close to the Castle.

The official guidebook to Conway Castle states "Where there might have been indiscriminate destruction, posterity can be grateful to the railway builders of a century ago for their skilfull and considerate handling of the problem which the presence of the Castle and walls at that particular spot posed for them".

Robert Stephenson's railway bridge is not pretty but it was a great step forward in engineering and accelerated the science of strength of materials, as well as being a triumph of man's ability to overcome great and novel engineering problems, in the same way as some of their descendants are tackling the great engineering problems posed by the North Sea Oilfields today.

THE RAILWAY BUILDERS

It must have been quite an experience for the middle class William Heap to join a firm in the railway business, because the navvies were the roughest, toughest, drunken, hardworking group of men ever seen.

The Victorian middle class engineers were usually men of integrity, with strict standards and usually of gentle upbringing, yet they controlled the toughest workforce the world has ever known. Surprisingly enough, there was mutual respect on both sides and they worked very well together.

At this time Thomas Carlyle wrote of the railway builders: "The country is greatly in a state of derangement, the harvest with its black potato fields, no great things, and all roads and lanes overrun with drunken navvies; for our great Caledonian Railway passes in this direction and all the world here, as elsewhere, calculates on getting to heaven by steam. I have not in all my travels seen anything uglier than that disorganic mass of labourers, sunk threefold in brutality by the threefold wages they are getting. The Yorkshire and Lancashire men, I hear, are reckoned to be the worst; and not without glad surprise I find the Irish are the best in point of behaviour. The Postman tells me that several of the poor Irish do regularly apply to him for money drafts, send their earnings home. The English, who eat twice as much beef, consume their residue on whisky, and do not trouble the postman."

When the Chester–Holyhead line reached Penmaenmawr in May 1846, three hundred Welshmen fought a pitched battle with the Irish navvies, the Riot Act was read by Magistrates and the military were called out—a not unusual occurrence.

Terry Coleman's book "The Railway Navvies", deals in depth with this splendid army of roughnecks who could outdrink, outfight and outwork anybody, and frequently did all over the world. Some idea of their drinking habits can be gauged by the fact that when Sir Morton Peto decided that heavy drinking had to stop during working hours he issued instructions to this effect but added, "A man has a right to bring a gallon with him if he likes in the morning". The English navvies ate two pounds of beef a day and drank gallons of beer and whisky, but they regularly shifted 20 tons of rock and earth and they built the world's railways with pick, shovel and wheelbarrow.

CONWAY RAILWAY BRIDGE

The Conway and Britannia bridges are landmarks in the history of engineering, not only were they the first large tubular bridges ever made, but for the first time the engineering problems posed were solved by a completely new science called "Strength of Materials", which first saw the light of day in William Fairbairn's famous Canal Street Works, Ancoats, Manchester.

The development of bridge design was given a great impetus by the railways because before the Liverpool to Manchester line was opened in 1830, there were only 20,000 bridges in the United Kingdom, but in the next twenty-five years there were no less than 25,000 bridges built by the railway engineers and William Heap played an important part in the building of the most famous of all.

When George Stephenson first started building the Liverpool to Manchester line, cast iron was used instead of stone, the traditional material, and the early iron bridges were arched following traditional bridge design, for example the best known of all those early bridges, that at Ironbridge in Shropshire, imitated the traditional designs.

It was during the building of the Liverpool–Manchester railway that Stephenson decided to break away from the arched design and he consulted Eaton Hodgkinson, a brilliant engineer who was born in the small country town of Anderton, near Northwich, Cheshire. Later, Hodgkinson worked with Sir William Fairbairn in establishing the principals of structural design which were the foundation of modern strength of materials.

Eaton Hodgkinson worked out the design of a cast iron beam and one of the first of these bridges ever made was over Water Street, Manchester, carrying the railway lines out of Liverpool Road Station.

George and Robert Stephenson were forceful and extremely good at solving problems by improvisation but they were working at the frontiers of engineering knowledge and the risks they took in accepting contracts before they had any inkling as to how they were going to solve the engineering difficulties are astonishing. Just as the great Victorian explorers marched off into the unknown, not knowing what they would find, so did the Victorian engineers. The result was a rapid acceleration in engineering knowledge, and a few mistakes.

The Chester to Holyhead railway was a typical example, because Robert Stephenson accepted the commission to build the railway before he had any idea as to how he was going to solve the two greatest engineering problems on the line, how to bridge the notorious, fast flowing. Menai Straits and the river Conway.

Stephenson's first thought was to build a wrought iron arch over the Straits but the Admiralty would not allow this because of the obstruction to shipping. He had designed a small tubular bridge by rule of thumb methods, which spanned the river Lea at Ware on the North East Coast in 1841, but he did not seriously consider this design until fate stepped in with one of those extraordinary co-incidences that it sometimes favours the brave and courageous.

At the Blackwall yard of Miller & Ravenhill on the river Thames, an iron steamship 180′ long by the name of "Prince of Wales", was being built. As she was being launched a bolt holding a cleat sheared and the bilge of the "Prince of Wales" came down on the wharf and the stern tipped downwards. She could not be freed for some hours and by then the tide had ebbed and this left the ship suspended for over 110′ at an angle with one end on the wharf and the other in the water. The onlookers expected that at any time the vessel would crack up but to their amazement when she was eventually freed by screw jacks and floated off on the next tide she had not suffered any structural damage apart from one or two damaged plates where she had hit the wharf.

** The Prince of Wales iron steam-vessel.*

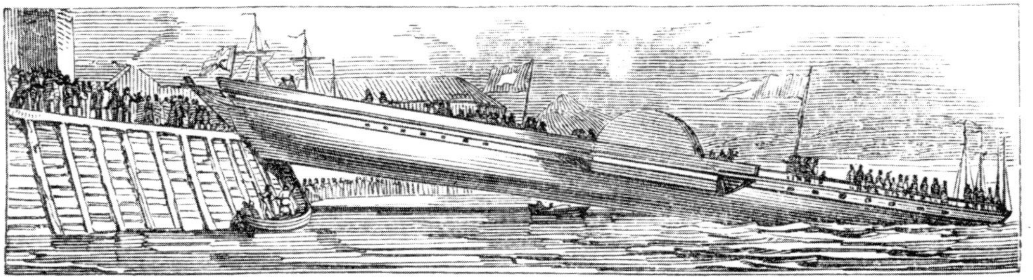

The accident to the "Prince of Wales". When Robert Stephenson was told that the vessel has not cracked up even though she was only supported at the two ends, it solved his problem of how to build the bridges over the Conway and Menai Straits. He would build iron "ships", float them in position, lift them and use them as tubular bridges.

This accident was related to Robert Stephenson and he immediately saw this as the answer to his problem, if the "Prince of Wales" could be suspended for 110′ and suffer no damage, then he would build his bridges as tubes on land and then float them in position. In fact that is what he did, he built iron "ships", lifted them in position and used them as bridges: but engineering knowledge and science had to be developed by others before his dream could be accomplished.

Although fate had given him the idea, he was still not clear about the precise design; his first thought was to have a circular tube, then he changed this to an elliptical shape, but it was obvious that further studies were needed and so he went to see William Fairbairn in Manchester.

SIR WILLIAM FAIRBAIRN'S CANAL STREET WORKS, ANCOATS, MANCHESTER

William Fairbairn (1789–1874) was born in Roxboroughshire, Scotland, but settled in Manchester in 1812 when he was twenty-three years of age and five years later he set up his own business with a workmate, Mr James Lillie. They rapidly built up Canal Street works into one of the most important in the country and not only did they manufacture water wheels, bridges and locomotives, but William Fairbairn set out to test engineering materials and structures in a most methodical way and developed the science of testing in a way never known before. Fairbairn tested rivetted joints and devised the tables still used today for this type of joint, he tested iron beams, trellis girders, devised creep tests, worked out and tested materials used in pressure vessels, worked on the properties of steam and for the first time determined directly the density of steam over a wide range of pressure and temperature.

Such was the quality of work done in the Canal Street works that engineers from all over Europe flocked there for training, one was Albert Escher son of the founder of the Zurich factory, another was the brother of Lord Kelvin who became Professor of Engineering at Glasgow University.

FINAL DESIGN OF TUBES

After carrying out tests William Fairbairn rejected Stephenson's ideas for a circular or elliptical tube and calculated that the most suitable structure would be a rectangular tube with a cellular base and roof for added strength.

For Stephenson's convenience they built at Fairbairn's Millwall works a one sixth scale model 80′ long, 4′–6′ deep and 2′ 8″ wide with supports 75′ apart. This was then loaded with weights and the deflection measured. It was found that the tubular construction enabled the bridge to support eleven times its own weight. Other tests were then carried out at Canal Street.

William Fairbairn and his men did their work well, for in those days trains were similar to "The Rocket" and yet today, one hundred and twenty eight years later, the "Irish Mail" and other great express trains thunder across the original bridges whistling their salute to the Victorian Engineers.

WILLIAM EVANS OF CAMBRIDGE

The contract for the masonry for Conway Bridge was let to William Evans of Cambridge, and the first stone was laid in the N.E. corner of the Conway tower on the 12th May 1846 by A. M. Ross the Superintendant Engineer for the Chester–Holyhead Railway.

The enquiry for the wrought iron work was sent to a number of companies in England and Scotland including William Fairbairn who also wanted to build the bridge, Hick & Co of Bolton (now Hick Hargreaves), Rothwell & Co of Bolton (still in existence) and Fairbairn's old partner James Lillie of Manchester.

None of the tenders were satisfactory and to some surprise William Evans offered to do the iron work as well as the masonry for the bridge. Objections were raised because he did not have a workshop for ironwork, but he said that he thought it was preferable to have the workshop there on the site in Conway and that if appointed he would do just that.

Robert Stephenson and Edwin Clark the Resident Engineer were both impressed with William Evans and so they gave him the enquiry and he was awarded the contract in October 1846 for the ironwork, for £145,190 and that price was to include for the fabrication of the six pontoons that were to be needed for floating the tubes in position and also for lifting the tubes in position.

Obviously, William Evans was not a man lacking in courage and enterprise.

As soon as his tender was accepted William Evans advertised for engineers and one of those taken on was William Heap then a young man twenty years old. He moved into lodgings in Conway and stayed there until the bridge was finished, taking part in the fabrication and lifting of a bridge that is a milestone in engineering history, where he met Robert Stephenson, Isambard Kingdom Brunel, Sir William Fairbairn, George Stephenson and for the first time, Thomas Brassey who was building 35 miles of the railway, but not the bridges.

PLANS FOR LIFTING THE TUBES

Stephenson instructed Edwin Clark to build the tubes partly on dry land 600′ from the site of the bridge, and partly on the pontoons which were in the river, but when the tubes were finished they would be floated off on the high tide, positioned, and then lifted bodily by "some means or other" into their final position. Stephenson was still very vague as to how they would carry out the daunting prospect of lifting the tubes weighing 1,300 tons and he made one or two suggestions which Edwin Clark dismissed as being impractical; and off he went back to London leaving Clark to work things out.

The method of lifting such enormous loads troubled Edwin Clark and Fairbairn but one day fate again came to the rescue, because on the 14th July 1846 Stephenson sent Clark to Manchester, and he had to change trains at Crewe. While he was waiting for his connection he saw some workmen lifting a very heavy cast iron water tank into position by means of a screw jack and to prevent accidents they packed up underneath as they lifted. This gave Clark the idea of lifting the tubes by means of hydraulic power through high pressure pumps driven by steam engines and, as with the water tank, they would pack up underneath as they lifted in case anything went wrong with the engine pumps or lifting "chains".

This was three months after work had started on the masonry towers which are higher than they need to have been because at one time they thought they would have to carry chains over the top of the towers to hoist the tubes.

The Roberts Jacquard Press, the world's first punched card controlled machine tool. It could automatically punch up to thirteen holes $1\frac{1}{8}$″ diameter in $\frac{3}{4}$″ plate at one time and the punches in operation were controlled by the punched card. William Heap had charge of this machine from its invention in 1847 until he set up on his own in 1866. The problem with this magnificent machine tool, at least a century ahead of its time, was to find enough work for it to do. ▶

PREPARING THE PLATES

Two platforms were built part on land, part on piers just above high water mark and on these platforms the tubes were constructed. The tubes were fabricated from $\frac{3}{4}''$ thick wrought iron plates 4′ to 8′ long and 2′ wide. The plates were rolled in Staffordshire and were transported to Runcorn by rail and then by road and sea to Conway. When they were delivered the plates were bowed and the first job was to straighten them. This was done by teams of blacksmiths with sledge hammers and a six inch thick flat bedplate on which the plates were hammered. The noise of this hammering was deafening and under the terrific hammering the flat bedplate eventually became concave so it was turned over and the hammering started again; this had to be done several times.

WORKSHOPS AT CONWAY

William Evans built a complete workshop at Conway equipped as follows:

A drawing office with two draughtsmen
1—20-hp steam engine with shafting to drive the machinery
3—Punching and shearing machines
1—Vertical drill
1—Lathe
3—Flypresses for punching holes
2—Travelling cranes and several fan blowers for the rivet furnaces. Air was carried from these blowers to furnaces in all parts of the works by square wooden ducting.

As well as the workshop machinery there were the steam engines and hydraulic pumps that were to be used to lift the bridge in position.

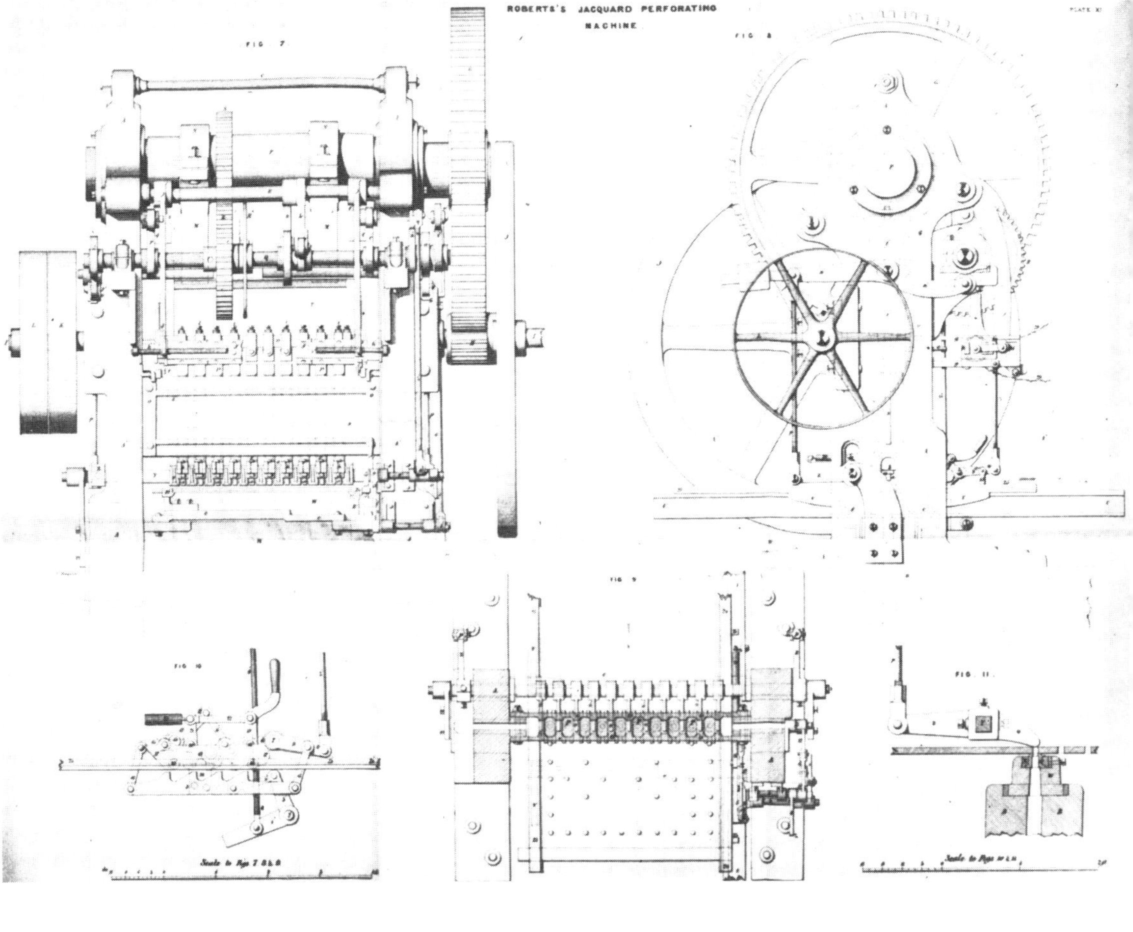

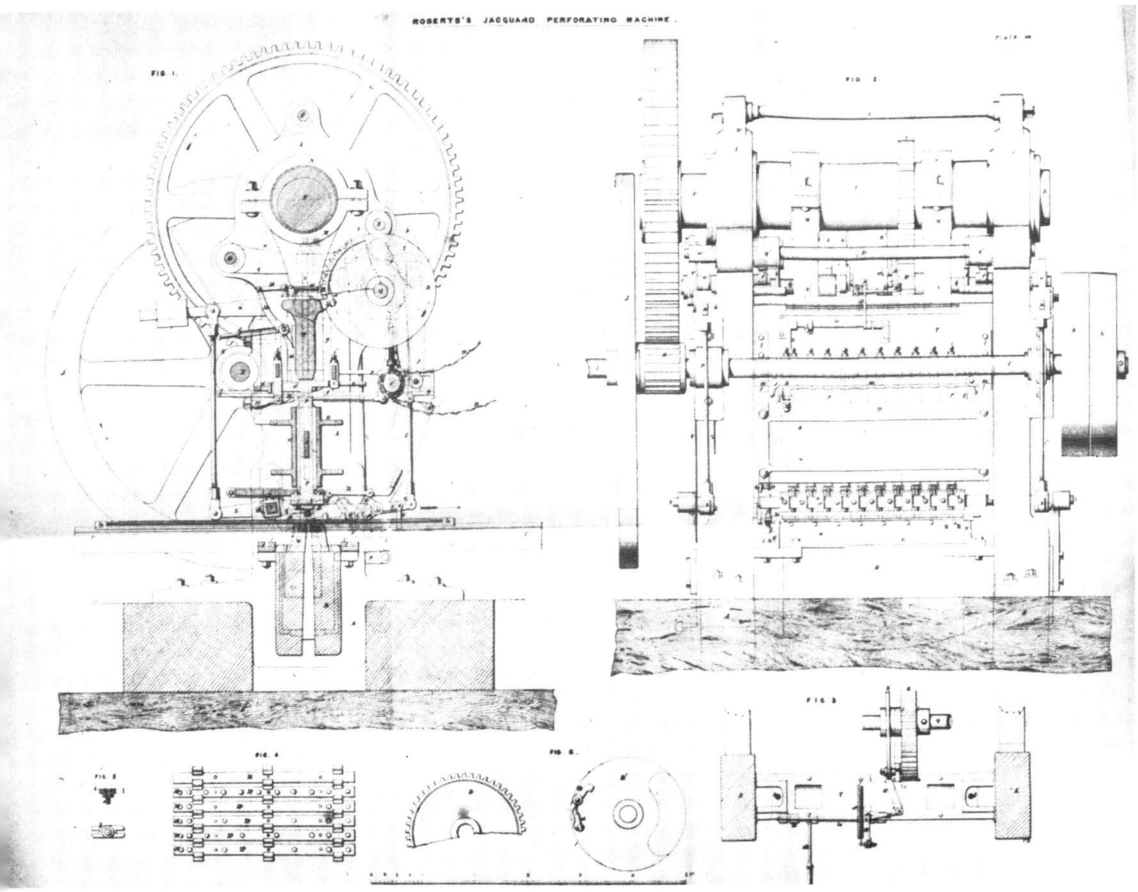

Roberts's Jacquard Perforating machine. The worlds first automatic punched "card" controlled press tool built in a matter of months by Richard Roberts of Manchester.

William Evans was an imaginative, enterprising man and he thought that they could speed up the punching of the plates if the Roberts' presses could be automated. The fly presses could punch one, two or three holes in the $\frac{3}{4}$ inch plate and a good punch lasted three days, and punched some 8,500 holes. However, Mr Evans was more ambitious and so he went to Manchester to see Richard Roberts, the famous engineer, to put to him some sketches for a multi-punch press. He went to the right man because Mr Roberts was also interested in cotton machinery as well as machine tools.

William Evans proposed that the new press should be automated by using the Jacquard card control which was being used in weaving machinery—the Jacquard loom, for lifting the warp and weft in a predetermined manner. The idea was to control the number of punches being used at any one time by the card control, and those punches not required were lifted, so that they did not touch the plates. Richard Roberts and his men set to work and in a remarkably short space of time, they had designed a completely new machine, 16′ high, 12′ wide and 12′ long with a flywheel 10′ diameter and main gear 12′ diameter with a 12″ diameter shaft, and this could punch up to thirteen holes $1\frac{1}{8}$″ diameter in $\frac{3}{4}$″ thick plate at one time and where the punch selection was by card exactly as proposed by William Evans.

This was the first time that an engineering machine tool was controlled by punched card and it worked superbly well.

Two men were employed in adjusting the plate to the sliding table which passed underneath the punches and then after all the holes had been punched two other men lifted the heavy plates off the machine. In this manner, the bottom plates of the tube which needed 144 holes, $1\frac{1}{8}$″ diameter punching in them were punched in under three minutes and this remarkable machine operated for long periods at the rate of 3,168 holes (twenty-two punched plates) an hour.

One can only marvel at the speed with which this sophisticated machine, even by modern standards, was developed and made in a matter of weeks by Mr Roberts and his men in Manchester.

William Heap had a long association with this machine because after the Conway bridge was finished he took it over to Drogheda in Ireland but due to problems dealt with later it was bought by Thomas Brassey and installed at Canada Works, Birkenhead, where it punched the holes in the plates for the Boyne Viaduct, the Victoria Bridge, the longest bridge in the world at that time, and later the Jumna River bridge in India. It was so good that the problem in later years was to find enough large jobs to use it on, because it operated so rapidly.

Surely the banks of the River Conway is the most improbable setting for the worlds first punched card, or tape, operated automatic multi-head press.

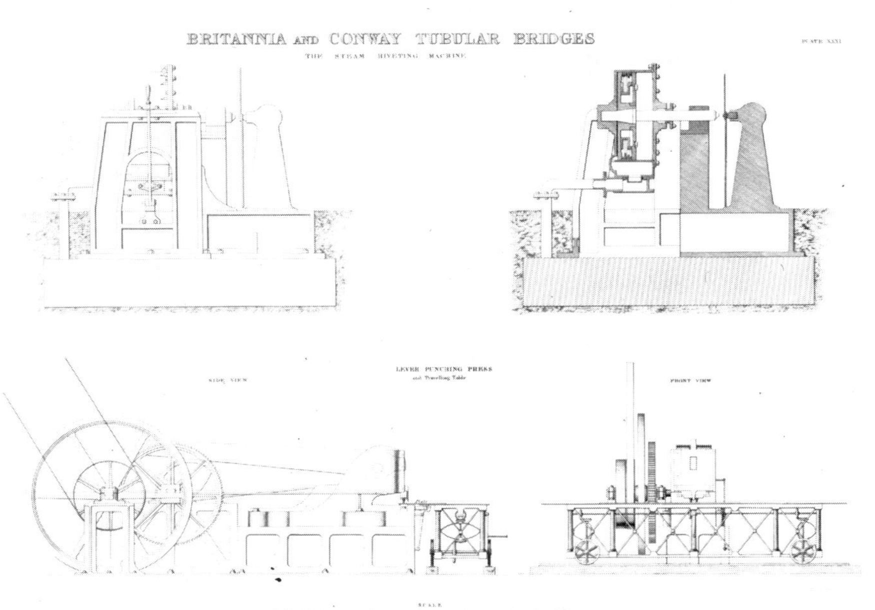

Riveting Machine used on Conway Tubular Bridge.

RIVETTING

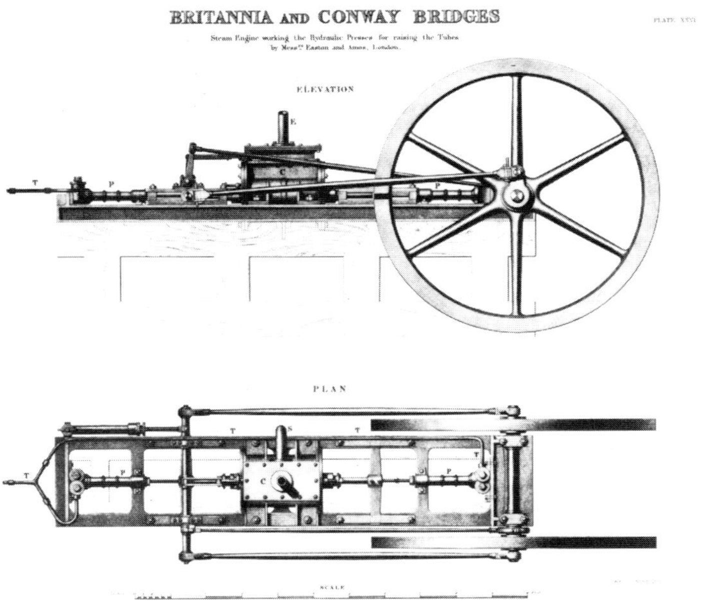

Combined Steam Engine and Pump, Conway Bridge.

A rivetting team comprised two rivetters, one "holder up" and two boys who were usually eleven or twelve years old. One boy worked on the portable forge which was fitted with a bellows, or he used a reverberating furnace to heat the rivets which varied from 2·25″ to 5″ long and they weighed $\frac{3}{4}$lb. When the rivets were red hot the boy would pick them up in his tongs and throw them in a curved arc some thirty or forty feet where they would be caught by the tongs of the second boy. He then put the rivet 1·5″ or 2″ into the hole. The "holder" would be inside with a heavy hammer and the two rivetters outside. After rivetting the rivets were then the finished off with the familiar dome shaped head.

Initially the rivetters started off with $2\frac{1}{2}$ lb to 4 lb hammers such as were used in boiler work but it was found that these were too light and they were changed to 7 lb hammers. The hammers used by the "holders" weighed between 30 lbs to 60 lbs. There was some difficulty in persuading the rivetters to change to the heavier hammers but once they started to use them they encountered no difficulties.

An eye witness described how people came from miles around to see the rivetters at work and the precision with which the eleven and twelve year old boys threw the rivets excited great admiration.

"By night red hot rivets ascending in graceful curves to a height of thirty or forty feet formed an interesting sight."

A steam rivetting machine made by Garforth of Dukinfield near Manchester, proved to be most successful and it was used wherever possible. Later this was taken to the Britannia Bridge over the Menai Straits, but there the contractors used London rivetters and they would not allow the machine to be used because they felt it would rob them of their craft.

The platform where the tubes were fabricated were made of wood and fire was a constant hazard, because of the presence of the furnaces and the red hot rivets flying through the air. To overcome this problem two fire engines were in constant attendance and cast iron fire mains were laid to every point of the platform and flexible hoses were available for the fire fighters. This system was later used on the Britannia platforms.

SIZE OF EACH TUBE

Each tube measures 412′ long, 14′ wide, 25′ high in the middle and 22′–3″ high at each end and the weight of each of the Conway tubes is 1,300 tons.

The expansion of the tubes due to temperature changes, was accommodated by twenty-four pairs of rollers and twelve 6″ diameter gun metal balls. Mr Clark suggested that they should attach to the end of one tube a system of levers and a dial, so that the relationship between temperature and movement could be measured, providing a means whereby they could learn more about the movement of large masses.

Eventually in February 1848, the first tube was finished and was ready to be towed into position, an eye witness recorded:

"When completed and resting on its massive platform, with crowds of busy workmen, the fuming chimney, the vast pontoons, all contributed to make the scene one of the most interesting and anamalous that was ever witnessed; especially when the peculiarity of the situation is remembered: the calm river floating idly by and the old castle, the work of hands long since crumbled to dust and of instruments long since eaten to rust, looking as it were in astonishment on the whole, while a crowd of Welsh country folk incessantly gaped with amazement at the idea of putting a long iron chest over their ancient river."

METHOD OF LIFTING THE BRIDGE

As we have already seen the idea of positioning the tubes was to float them into position and then to slowly hoist them into place. Robert Stephenson was not very happy about floating the tubes into position but Edwin Clark was very persuasive and finally convinced Stephenson

Conway 1848— Working on the second tube.

that it could be done with proper planning and with the right man in charge of the marine operation. The man chosen for this task was Captain Claxon R.N., who had achieved fame by refloating Brunel's superb ship "The Great Britain", the mightiest ship ever built at that time, when she ran aground in Dundrum Bay, Ireland two years earlier on a voyage from Liverpool to New York.

The hydraulic machinery for lifting the bridge was made by Easton & Amos of Grove Iron Works, Southwark, London, and they positioned at the top of each tower two large hydraulic rams, 18″ inside diameter and 36″ outside diameter, giving a wall-thickness of 9″. To the top of each ram a cross-head was secured two square feet thick, with two square holes through which the chain passed. These "chains" were made of flat wrought iron bars $1\frac{1}{2}$″ thick, 7″ wide and 6′ long. Each ram lifted two chains comprised of nine links, containing eight bars in the upper links, but only four in the lower. The stroke of the ram was 6′ and at the end of each stroke the bridge was to be supported, the chains disconnected, one link taken out to shorten it, the ram lowered and the chain reconnected, and then the lift started again.

HYDRAULIC PUMP

In the recesses where the bridge was to be finally positioned were two steam pumps or in the words of an eye witness "steam engines of a peculiar construction". The engines were horizontal, high pressure (for those days) engines, and the piston rod ran through the cylinder at both ends. The drive end was connected to a con-rod and flywheels and the other end formed the piston of the hydraulic pump.

From the pump ran a $\frac{3}{8}$″ bore wrought iron tube and this conveyed the hydraulic pressure to the lifting cylinders.

Details of the pumps and hydraulic rams are:

Steam pressure	50 p.s.i.g.
Engine bore	12″
Stroke	16″
Pump Piston diameter	$1\frac{1}{16}$″
Pump speed	40 strokes/minute
Pump discharge pressure	8,556 p.s.i.g. (3 ton/sq in)
Hydraulic ram O.D.	36″
Hydraulic ram I.D.	20″ (after modification)

Great difficulty was experienced when the rams were being constructed because such high pressures had never been used before. When the ram was first tested it was found that despite the 9″-thick walls the bore increased to such an extent that power was lost. A new larger piston was made but the same thing happened, a third piston was then made with the same result and the engineers were despairing when somebody noticed that although the bore was still increasing, the rate of increase was dropping and that furthermore, the outside diameter of the cylinder had not increased. This indicated that the wrought iron was being compressed under high pressure and so they carried on until compression was completed and the cylinder bore did not change under pressure.

When finished the rams were capable of exerting a lift of 900 tons each which was adequate for Conway as well as the heavier Britannia tubes.

POSITIONING THE TUBES

On Monday 6th March 1848, twelve months after construction of the tubes had started the great experiment was made. Some of the greatest engineers in the world at that time journeyed to North Wales to witness the lifting of the tubes. Robert Stephenson, the engineer for the Chester–Holyhead line, his father George, Isambard Kingdom Brunel, William Fairbairn of Manchester, Eaton Hodgkinson and all the gentry for miles around came with a great crowd of locals.

Conway 1848—Preparing for the second lift.

Conway. Floating the second tube into position 12th October 1848.

The first tube with its ends blanked off to make it watertight rested on the platforms with the pontoons under them; as the tide started to rise, the six immense pontoons each 100′ long were floated underneath by sailors from Liverpool under the command of Captain Claxon; as the tide continued to rise the pontoons lifted the bridge and it was ready to be floated into position. There were seventy men on the pumps, valves and crabs and two hundred sailors on the capstans and pontoons with several rescue boats in attendance. Captain Claxon with his megaphone and a system of signals directed the bridge across the river towards the estuary until eventually it was in position, resting on two temporary stone piers. This was a very tricky operation because this was by far the largest iron "vessel" ever to float at that time, and to manoeuver such an ungainly craft into position on the fast flowing Conway was no easy task. As soon as it was accomplished, great cheers rang out from the onlookers and workmen and in his exuberance Captain Claxon smashed his megaphone and threw it into the river.

When the first tube was in position ready for the great lift, a restless Robert Stephenson, a concerned Edwin Clark, a cigar smoking Brunel and a cadavorous William Fairbairn, climbed on to the top of the tube and witnessed the chains being secured to each end of the monster structure. The boilers were then fired up to full pressure, the engines and lifting rams were manned and the excitement was intense—many wondered if it would be possible to lift such a gigantic load by means of the tiny $\frac{3}{8}''$ bore pipes and even Edwin Clark was extremely tense and concerned. An eye witness reported:

"Regarded in itself this little $\frac{3}{8}''$ bore tube was the least imposing portion of the whole mechanism and no one who looked at it by the side of the vastly proportional instrument it was attached to would have believed that this tiny cylinder was the channel of a force equalling 700 or 800 tons. Could it be possible that this vast work was to be lifted by the direct instrumentality of two tubes the size of a quill barrel?"

Conway Tubular Bridge.

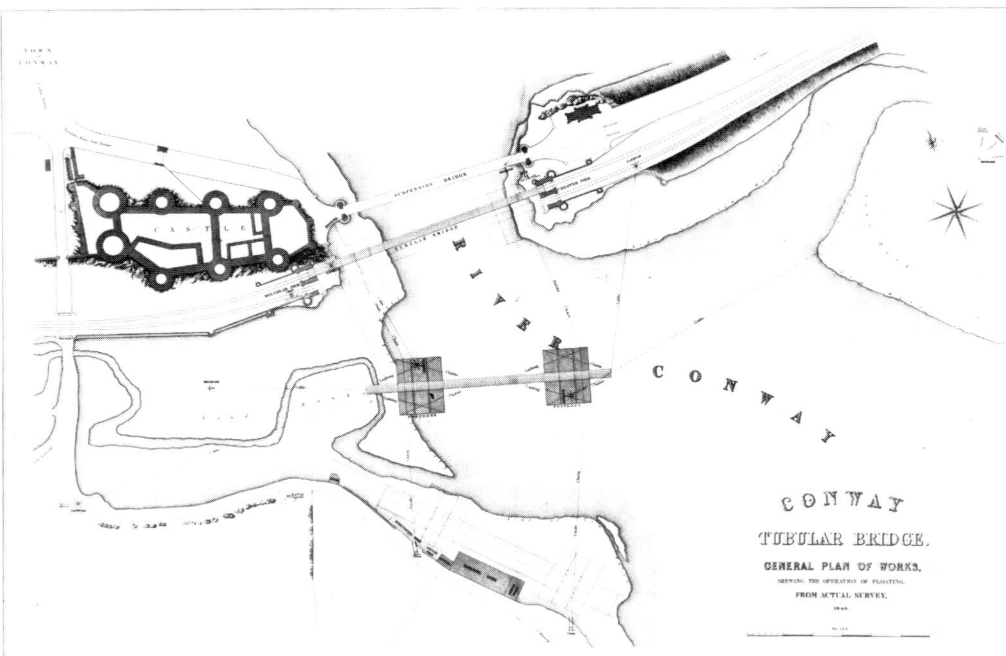

When all was ready Edwin Clark gave the signal and the engines at both ends of the bridge started work and slowly the mighty structure started to lift, but the bridge started a tremulous motion like a wave running along its length. The lift was stopped and the engineers consulted together and concluded that the motion was being created by both engines working together at the same time. The lift was started again but this time with each engine working alternatively.

Under the bridge were slung two platforms at both ends and on these were teams of men who put timber packings under the tube every inch, so that if any of the lifting gear failed, the tube could not fall. Following this team came the stonemasons who built the final stone supports of white limestone from Penmaemawr and the Great Orme. The stone used for the interior of the columns was sandstone from Runcorn.

An eye witness wrote:

"Amid the buzz of voices, the muffled sound of clacking valves and the hurrying too and fro of swarthy mechanics, the Tubular Bridge rose majestically but with great slowness, into the air. At every rise of six feet the engines were stopped and the chains readjusted to the head of the ram and the top links removed. By a succession of such rises the tubes finally reached its desired elevation of about 24 ft, and there dangled in the air as though a mere plaything in the hands of two hydraulic giants."

The masonry was built up underneath and then the bridge was allowed to take up its final position. The anxiety of William Evans, his men, Stephenson, Clark and the other great engineers present, was at an end and the tubular bridge lay across the river Conway, a great and lasting monument to the skill and courage of the Victorian engineers who had worked so hard to build it.

Railway lines were then laid through the tunnel and a locomotive and carriages loaded with 300 tons of ballast were rolled on to the bridge and left for three days, but the maximum deflection did not exceed $1\frac{1}{2}''$ and on removal of the load the deflection disappeared.

So a new type of scientifically designed and tested bridge was born and a novel method of lifting had been invented and proved, the sceptics were confounded and another link in the nation's network of railways was forged by sweat and genius.

The "London Illustrated News" later reported:

"Since its erection the bridge has been continuously used and the vast hollow, which a few months ago resounded with the deafening clatter of the rivetters' hammers, now roars with the rush of carriages and re-echoes in a voice like thunder the hoarse and impetuous expectations of the flying locomotive. The mathematicians still nurse their forebodings; but may God forbid that a work of so much skill and ingenuity, and the destruction of which would inevitably involve so fearful a loss of life, should become a mass of ruins! We do not share these fears, experiment has long since settled the question; and we believe that nothing but some anomalous and unforseen class of circumstances could injure the security of the Tube Bridge. The Tube Bridge is pre-eminently a work of our own era; it is one of those vast and complicated efforts of skill which no previous period of the world's history could command."

The young William Heap had to work extremely hard from very early in the morning until late at night, but they were heady days for a young engineer and he knew that he was taking part in a great and exciting adventure.

The second tube was floated on the 12th October 1848 and was raised on the 30th October 1848, this tube was loaded with 235 tons–14 cwt–2 qrs and it deflected 1.86″ and again this disappeared when the load was removed. When a normal train was run over the deflection was less than $\frac{1}{8}''$.

William Fairbairn patented the tubular bridge and there were some hard feelings between him and Stephenson because Fairbairn felt that Stephenson had not given him due credit for the work he had done in perfecting the design, this was a pity but understandable.

PUBLIC DINNER

On the 17th May 1848 a public dinner was given by the Gentry of Conway to Mr Robert Stephenson to celebrate the successful erection of the tubular bridge. This dinner was held in a temporary pavilion at the back of the Castle Hotel, Conway. The tent was 70′ long by 25′ wide and at each end of the pavilion were the gigantic initials "R. S." in a blaze of variegated lamps and the whole was festooned with flowers and evergreen. The interior of the tent was lined with crimson and white drapes and formed a splendid setting for the celebration.

One hundred and thirty tickets were sold at one guinea each and young William Heap found himself in exalted company because the guests included:

Hon. Edward Mostyn
Lloyd Mostyn M.P.
Robert Stephenson
George Stephenson
Captain Claxon
Wm. Evans

Mr Fairbairn and Mr Brunel sent letters of apology.

At the head table sat the Hon. Edward Mostyn, Lloyd Mostyn M.P., Robert Stephenson sat on his right and George on his left. The food was superb and the vintage wine flowed freely.

Robert Stephenson spoke about the bridge and one speaker marvelled at the speed of the railway "We are now whirled along at a speed exceeding that of the fastest horse".

Messrs Ryalls, G. Holden Jnr., Roberts and G. Holden of Liverpool were the Glee Singers and the latter gentleman officiated as conductor of the piano-forte, and they entertained the distinguished guests with renderings of:

"The Winds Whistle Cold"
"The Huge, Huge, Globe"
"Mynheer Van Dunk"

There was great pride in the skill of British engineers in building Conway Bridge and "Chambers Journal" reported:

"There are men who are in raptures with the engineering skill which reared the Pyramids, built Baalbec and adorned Petra, but turn with a smile of pity to the 'puny efforts', as they call them of modern times. If the eye of such persons rests on this page let them accompany us while we describe one of the most surprising and stupendous efforts of modern engineering enterprise—the Tubular Bridge—and they will become acquainted with a work which Egypt and the ancient might have been proud of, but could never have excelled."

The Conway Tubular Bridge today. Although it was only built for locomotives similar to the Rocket, it now carries modern trains such as the one leaving the bridge. The depression in the foreshore on the right is where the first tube was made.

THE BRITANNIA BRIDGE, MENAI STRAITS, NORTH WALES

Construction of tubes on the Carnarvon Shore.

Following the successful construction and lifting of the Conway Bridge came the building of the Britannia Bridge across the Menai Straits. It was a great consolation to the engineers to know that they now had a tried and proven method of making and positioning the tubes, but although the problems were the same as Conway they were of a greater magnitude. There were four spans each of two tubes, not one double, as at Conway; the Straits are very difficult waters to navigate as any yachtsman will tell; the lift of the tubes was one hundred feet not 24′ and each tube was 472′ long—72′ longer than Conway and weighed 1,600 tons as against Conway's 1,300 tons.

William Evans was not the Contractor for the Britannia Bridge but in view of the experience that the Evans' men had gained at Conway, they were frequently consulted and their lifting gear and pontoons were used as the two sites are only 17 miles apart. William Heap was one of those who periodically visited and worked at Menai and he was able to see the progress of the Britannia Bridge.

Two complete villages were built on either side of the Straits near to the site for the bridge, including huts for the men, shops, schools for the children and a chapel of worship. There were some complaints about the prices of the goods in the shops but in general the villages were almost self supporting.

At Conway, Stephenson built his bridge alongside Telford's suspension bridge and similarly at Menai the tubular bridge can be seen from Telford's delicate and elegant road bridge.

Apparatus for lifting the tube of the Britannia bridge.

The Britannia tubes were constructed on the bank some 600′ from the permanent site and as well as the Conway pontoons the contractors built several large ones 98′ long by 5′ wide and 11′ deep drawing 5′ when fully loaded. In the bottom of the pontoons were large valves which were normally open to admit the tide, but when they were ready to lift, the valves were closed and the water pumped out.

Captain Claxon was again in charge of the floating of the tubes, each tube was 472′ long and 98′ high and some idea of the navigational problems may be gained by noting that each tube was three times as long as the largest ship then afloat anywhere in the world at that time—the "Great Britain".

Large pontoons with giant capstans bolted to them were towed out into mid-stream and anchored there. Each capstan was worked by fifty men, hawsers passed around the capstans from the shore and the masonry tower. There were no less than 600 men employed in the floating and positioning of the tubes and of this number, three hundred and eighty six were sailors from Liverpool.

Because of the distances involved between the various teams of men Captain Claxon evolved a system of signalling by means of boards with numbers and letters painted on.

The towers were built of Anglesey marble with an interior of Runcorn sandstone as are those at Conway. The blocks of Anglesey marble are gigantic and each block weighs between 10 and 12 tons and conveying them and lifting them into position to build the towers in mid-stream was no mean feat. At the top of the piers there are rectangular slots which were built in in case it was necessary to have chain supports for the tubes at a later date, but of course these were not needed.

STEAM ENGINES AND LIFTING TACKLE

There were two horizontal steam engines, made of a similar design to those used at Conway, each of 40-hp with cylinders 17″ I.D. × 16″ stroke. The pump piston was $1\frac{1}{16}$″ diameter and the wrought iron pipe used at Menai was 0·5″ internal diameter and 1″ outside diameter.

The hydraulic presses used for lifting were 20′ I.D. × 42″ O.D. and were made at the Bank Quay Foundry in Warrington. In his report on the Britannia Bridge Edwin Clark calculated that the pressure generated by the pump was 5,040 lb/circular inch, that is 6,400 p.s.i. which was 2,150 p.s.i. less than the Conway pumps.

ENGINEERS AND CONTRACTORS

Engineer—Robert Stephenson
Resident Engineer—Edwin Clark
Consultants—William Fairbairn
Masonry and scaffolding—Messrs Nowell, Hemmingway and Pearson
Fabrication of one tube—Messrs Garforth of Dukinfield
Remaining seven tubes – Messrs Ditchburn & C. Mare of Blackwall, London
Hydraulic presses—Easton & Amos of Southwark, London

Most of the platers were from London, the stonemasons and labourers were Welsh and the sailors were from Liverpool.

LIFT OF THE FIRST TUBE

On the 27th June 1849 the first tube was floated and prepared for lifting. The correspondent of the "Illustrated London News" gave this eyewitness account on the 30th June 1849:

"Mr Stephenson and other Engineers gathered to witness and direct the operation, as with Conway, Captain Claxon R.N. was in charge of the floating and positioning of the tubes and he

Britannia Bridge, Menai Straits.

Britannia Bridge, Meani Straits (Entrance to the Tubular Bridge).

used his speaking trumpet and he had a number of men with cards on which were printed different letters which indicated to the various capstans, so that no mistake could occur as to which capstan should be worked, and flags red, blue or white, signalled what particular movement should be made with each.

'About 7-30 in the evening the first perceptable motion which indicated that the tide was lifting the mass was observed, at Mr Stephenson's desire the depth of water was ascertained and the exact time noted. In a few minutes the motion was plainly visible; the tube being fairly moved forward some inches. This movement was one of intense interest; the huge bulk gliding as gently and easily forward as if she had been a small boat. The spectators seemed spellbound for no shouts or exclamations were heard, as all watched silently the silent course of the heavily freighted pontoons. The only sounds heared were the shouts from Captain Claxon as he gave directions to "let go ropes', to 'haul in faster' etc and broadside on the tube floated majestically into the centre of the stream.

'I then left my station and ran to the entrance to the works, where I got into a boat and bade the men to pull out as fast as they could into the middle of the Straits. This was no easy task as the tide was running strong; but it afforded me several splendid views of the floating mass and one was especially fine; the tube coming directly on down stream; the distant hills covered with trees; two or three small vessels and a steamer—its smoke blending well with the scene—forms a capital background; whilst on one side, in long stretching perspective stood the three unfinished tubes destined, ere long, to form with the one then speeding on its journey one grand and unique roadway.

"It was impossible to see this imposing sight and not feel its singleness, if we may so speak. Anything so mighty of its kind had never been seen before; again it would assuredly be; but it was like the first voyage made by the first steam vessel—something till then unique."

When the tube was eventually positioned it was lifted into its place in the same way as at Conway being lifted at the rate of 2″ per minute and again men worked on platforms slung under the tube following the tube up with wooden packing pieces, and they in turn were followed by the Welsh stonemasons.

ACCIDENT

On Friday 17th August 1849, after three of its lifts had been successfully accomplished on the previous days; the lift was proceeding and 2′–6″ of an individual lift of 6′ attained when between 11 am and noon the bottom of one of the cylinders of one of the long presses "burst out" and the 1·5 ton piece fell 70′ to 80′ on to the top of the tube below.

As already related Stephenson and Clark had insisted that the wooden packings, 1″ thick, should be inserted as rapidly as possible as the tube rose. These packings were then carefully removed piece by piece and the spaces filled in with brick work in cement so as to be nearly flush with the outer wall of the tower.

As it was the fall of the tube was only one inch but even so the 1,600 ton tube caused quite a shock. The bottom of the cylinder in falling unfortunately struck a sailor who was ascending a rope ladder from the tube to the hydraulic ram and killed him.

A new cylinder was made in six weeks by Bank Quay Foundry in Warrington and when it went into service this worked very well, all the other tubes were lifted successfully.

The Britannia Bridge took four years from work first starting on the masonry, the first stone was laid in May 1846, the first rivet was hammered in on the 10th August 1847 and the last 2,000,000 rivet was hammered in by Robert Stephenson himself and to this day this rivet is painted white; the first train went through in March 1850.

The cost of the Britannia Bridge was £601,865 and as Edwin Clark commented in his book on the Conway and Britannia Bridges only twenty years earlier these bridges could not have been built because "not only was there no machinery nor tools for the manufacture and working of such heavy plates, but that intelligent and valuable class of men who carry out such operations with an energy and skill in the practical detail and a sound mechanical appreciation of what is required, which it is a delight to witness was not then in existance: they are the peculiar offspring of railway enterprise and among the most valuable fruits of its harvest." One such man was William Heap.

After the bridge was finished the great hydraulic cylinders that had been used for lifting the tubes were exhibited at the Great Exhibition of 1851 held in the Crystal Palace, Hyde Park, to take stock of the products of the Industrial Revolution and of Victorian achievement.

THE FIRE IN 1970

A modern sequel to the story of the Britannia Bridge is that on the 23rd May 1970 two young boys entered the tube on the Anglesey side looking for bats, and they accidentally set fire to the timber lining of the tube. This badly damaged some of the plates before the fire was discovered such that the bridge had to be closed.

After studying the consultants reports it was decided that a new supporting arched structure should be built under the tubes and sometime in the future a road bridge is going to be built across the stone towers. While the Britannia Bridge was out of action, the island of Anglesey suffered great difficulties indicating the continuing importance of this great Victorian Bridge, one hundred and twenty-five years after it was built.

THE BOYNE VIADUCT, DROGHEDA, IRELAND

The next large contract William Heap worked on was also an historic bridge because it was one of the first large lattice bridges to be built of wrought iron and where the problem would prove to be so great that his employer William Evans would be forced into liquidation.

The lattice work bridge had been widely used in America because this design was ideal for building with timber which was found in abundance all over America. Sir John MacNeill was the Consultant Engineer for the Dublin and Drogheda Railway and in the early 1840's he had built the first ever lattice bridge in wrought iron to carry a private road across the line and having proved the design he used the lattice design to build a bridge across the Royal Canal.

In 1845 it was decided to join up the Dublin to Drogheda line to the Ulster Railway and so make a continuous line from Dublin to Belfast and the new system was to be re-christened the Dublin and Belfast Junction Railway.

The main engineering problem was to build a bridge across the Boyne at Drogheda and as at Menai the Admiralty insisted that the bridge must have adequate headroom for shipping and specified that the headroom should be 90′ and that there should be a clearway of 250′. Again as at Menai this ruled out an arched bridge and the choice lay between a lattice bridge and a tubular model. Sir John MacNeill sent out a general enquiry asking for suggested designs to several ironworks and he received tenders for two or three lattice work viaducts and a very comprehensive and detailed proposal from William Fairbairn in Manchester for a tubular bridge. Sir John had also made a very sketchy design of a lattice bridge himself and he showed this and Fairbairn's proposals to the Board.

The Boyne Viaduct, Drogheda, Ireland.

About this time (1849) James Barton joined the D. & B. J. R. as Engineer and when he saw Sir John's plans he was critical because they were incomplete and far too vague; so he seconded one of his own men, Mr A. Schaw, to work in Sir John MacNeill's Dublin office in order to complete the design. James Barton was a most thorough man and he checked and re-checked the lattice bridge design. Then he had a 60′ span made in Belfast and tested this to destruction. This revealed weaknesses in the original design so they started again and Barton and Schaw worked fifteen to twenty hours a day to complete the revised design.

In May 1851 the plans were sent out again for re-tender, and William Evans was awarded the contract at a price of £68,000 and William Heap moved to Drogheda to prepare for the production of the wrought iron which it was expected would be needed within a few months because the bridge had to be finished by 1853.

William Evans in his customary energetic style set to work and recruited labourers and stonemasons locally paying enhanced wages to the men preparing the foundations. Unhappily they ran into considerable trouble with the alluvial soil, particularly with the foundation for pier 14 which was to be positioned in the water and was to carry one of the iron spans across the Boyne.

The coffer dams were built and day after day the men worked hard digging out the subsoil but all they found were unstable clay, loose fragments of rock and small boulders. William Heap was kept occupied by building a temporary wooden arch across the Boyne which was to be the support for the latticed iron structure during construction, but month after month went by and still the men digging in the coffer dams did not hit a solid foundation.

Trouble was looming ahead for William Evans because the bridge had to be finished in time to carry traffic from the North down to Dublin for the Irish Great Industrial Exhibition of 1853 and daily it grew more and more obvious that in view of the problems with the foundation that the viaduct would not be finished in time.

The long delays were now creating financial problems for William Evans but he was determined not to let the Railway company down and so he mortgaged all his plant and tackle, offering to convert the sturdy, but temporary, wooden arch into a temporary railway bridge so that the trains could cross the river. His offer was greatfully accepted and £2,000 was spent on strengthening the wooden arch. A track was laid across it and the Dublin to Belfast line was complete, albeit temporarily.

On 11th May 1853 a test train of a locomotive with four wagons of ballast was steamed on to the temporary arch and Captain Wynne of the Board of Trade tested for deflection and found this was only $\frac{3}{8}''$ and so he gave permission for trains to cross the bridge at a speed of 4 m.p.h.

Work continued in the coffer dam and they were 30′ down but still had not found bottom, so William Evan's financial problems were now enormous. Although he fought on as long as he could, desparately hoping to gain enough time for his men to find the foundation he was seeking, he was within a few weeks of finishing the temporary structure when he was forced into liquidation and William Heap at the age of thiry-three found himself out of work—but not for long.

At the same time over the other side of the Irish Sea work had just started on a new works to make bridges, locomotives and other railway equipment. The man behind the new venture was Thomas Brassey, a contractor, whom William Heap had first met at Conway and a man with an excellent reputation as a contractor and employer. The prime reason for building the new works, called "Canada Works" was to build the equipment for the Grand Trunk Railway in Canada and Robert Stephenson was the engineer. The most difficult part of the contract was the building of the "Victoria Bridge", the largest bridge in the world and to this day the longest tubular bridge ever built.

Robert Stephenson had been very impressed with the work of William Evans and it is possible that he suggested to Thomas Brassey that William Evans and William Heap would be ideal for running the Bridge Department, but whatever the reason William Evans joined Thomas Brassey in 1853 as Manager of the Bridge Department and William Heap became sub-Manager.

Back in Ireland the Dublin and Belfast Junction Railway decided that they would finish the bridge themselves, because once they reached a firm foundation for Pier 14, the rest would be easy because of the wooden arch already spanning the Boyne.

James Barton set his men to work again in the coffer dam for pier 14 and in October 1853 they were rewarded by finding sure foundations at a depth 45 feet below the low water level of the river. William Evans had been so near success, had gambled, and had lost everything. After his splendid work on the Conway Bridge he deserved a kinder fate.

Thomas Brassey had bought the magnificent Roberts-Jacquard press from William Evans and this produced all the iron work for the Boyne Bridge and James Barton's men rivetted it together to form the first great wrought iron lattice work bridge. The punched cards for the Boyne, the Conway and the Victoria bridges were kept at Canada Works for many years as a reminder of the three great historic bridges that the machine helped to build.

William Evans did not stay long at Canada Works and in 1856 William Heap took over as Manager of the Bridge Department, a post he was to hold for ten years.

CHAPTER 3

Thomas Brassey and Canada Works

Thomas Brassey in 1862.

THOMAS BRASSEY—RAILWAY CONTRACTOR

William Heap's new employer was a man who was to exert a great influence on him and through William Heap on this company. After Thomas Brassey had died and his company went into liquidation we took over many of his connections and to this day we benefit from the work and influence of this Victorian giant.

Today Thomas Brassey is one of the great unsung heroes of the Victorian Age, a gentle, kindly man he nevertheless was the World's first great Contractor employing 85,000 men, he was respected by his competitors, loved by his men, a man of great integrity.

Thomas Brassey was born to John and Elizabeth Brassey on the 7th November 1805 at Buerton which is about five miles south east of Chester on his father's farm; and as well as their own farm they rented another neighbouring farm from the Marquis of Westminster. The family originated from Bulkeley which is near Malpas, which in turn is about fifteen miles south east of Chester, and although at the time Thomas was born they were living on the Buerton farm they owned, and the family still own, Bulkley Manor House and 400 acres of good Cheshire farmland which have been in the Brassey family for over six hundred years.

In 1805 the population of Buerton was only thirty-nine and the community comprised the Brassey farm and about fourteen scattered cottages. In those days there was no school in the village and so Thomas must have had some private tuition from his mother or a governess.

In 1817 when he was twelve, Thomas was sent to school in Chester and his parents chose that run by Mr Harling known as The Boarding School, St John's Rectory, Vicar's Lane, Chester. There were at that time about fifteen boarding schools in Chester patronised by families living on farms or country estates in the surrounding area. Although the Brassey farm was only five miles from the city the roads were poor and it was out of the question to expect a child to do this journey twice a day, so Thomas Brassey became a boarder.

Mr William Harling gave Thomas a sound basic education and although the young Brassey was not a brilliant scholar he was a hard worker, intelligent and showed a certain aptitude for mathematics. He was such a balanced, happy man that he no doubt enjoyed life in the busy little city, which must be one of the most perfect towns in Europe with its Roman road layout, virtually unchanged for 2,000 years, its unique "Rows"—the elevated covered pavement, its lovely black and white timbered buildings all encircled by the ancient city walls; the only complete city walls in Britain.

Such was the background of Thomas Brassey, Yeoman stock, with roots going back deep in the soil of Cheshire since the Norman Conquest. Some may scoff, but it matters this question of roots and soil—England has been particularly fortunate in the number of men who came from this sort of background to play an honourable part in our history.

In 1821, when Thomas Brassey was sixteen, he took his farewell of Mr Harling and was articled to Mr Lawton who was a Land Agent and Surveyor. Mr Lawton was agent for several people including Francis Richard Price of Overton near Wrexham. Mr Price was a landowner of considerable standing and part of his estate included the township of Birkenhead which at that time only comprised a few straggling houses on a bleak and sterile coast.

In 1824 Thomas Telford was building the A5 from Shrewsbury to Holyhead and he employed a surveyor from Oswestry called Mr Penson to do some of the surveying, but he had so much work to do that he asked Mr Lawton if he could borrow Thomas Brassey to help in the surveying. For two years young Brassey worked surveying the land for the A5 and he gained invaluable experience which was to stand him in good stead in later life.

In 1826 two events occurred which later had a considerable influence on us, firstly Mr Lawton realising that Birkenhead had considerable development potential sent Thomas Brassey there to open an office and almost fifty miles away, in Isles House in the shadow of Pendle Hill, a baby, later christened William, was born to Mary Heap on the 23rd September 1826.

In 1801 the population of Birkenhead was 110 people but by 1826 the town had grown to about 4,000 and in the next decade the size increased to 6,000. They were busy days for Thomas Brassey and in a short time Mr Lawton made him a partner.

Thomas Brassey then borrowed some money from his father and bought some clay bearing land in the dock area of Birkenhead and set up a brick works. One of his first large contracts was the supply of some millions of bricks for the new Customs House that was then being built over the river. In 1839 he took advantage of his work with Telford and he tendered for, and won, the contract for building the New Chester Road from Tranmere to Bromborough. It is a wide, well surfaced road and was built to accommodate the increased coach and carriage traffic from Birkenhead to Chester and it included a large bridge needed to span Bromborough Pool.

One of the first residents of Birkenhead was Joseph Harrison, who lived in a pretty cottage called Mersey Cottage right on the banks of the Mersey shore close to Monk's Ferry. He had two sons Henry and George and a daughter Maria, and Thomas was very friendly with the Harrison boys and they used to invite him home where he got to know Maria and found her a very pleasant young lady. Eventually he and Maria were married and they lived in a large house in Park Road South, Claughton, Birkenhead, which Thomas built himself.

Mr and Mrs Brassey were well liked in Birkenhead, as indeed they were later elsewhere, and the owner of the Birkenhead Brewery Company, Henry Kelsall Aspinall, knew Thomas Brassey very well and he described him as "a gentleman in the fullest sense of the word; quiet, simple living and simple hearted, amiable and kind to a degree." Every person who knew him, including William Heap, had the same opinion.

For the next few years Thomas Brassey carried on as Agent for Mr Price and he also carried out one or two more building contracts including the building of Holy Trinity Church, Birkenhead; he opened a lime works and a builders merchants' yard and gradually prospered.

On fine summer weekends Mr and Mrs Brassey enjoyed taking a pony and trap and going to West Kirby to sit on the beach and then have a simple, but excellent tea of shrimps and home made bread and butter. Years later when he was a millionaire several times over and was living in a fine house in Lowndes Square, London, he still remembered those teas. He had been given a magnificent banquet by the Emperor and Empress Louis Napoleon of France in the splendour of the Tuileries, he had been wined and dined by the Emperor of Austria, and was eagerly sought by half the nobility in London; but when one day he had to visit Canada Works in Birkenhead he wrote to Mr Aspinall saying that he would like to visit the Aspinalls in their summer residence in West Kirby as he had not been to West Kirby for years. When Henry Aspinall asked him if he would like a dinner party or tea Brassey wrote back; "Certainly, tea, shrimps and bread and butter, nothing else. We can get plenty of dinners in London." He came up and after he had finished his business in Canada Works and had seen his brother in law George Harrison, he spent the whole afternoon on the beach with the Aspinalls and then they retired for their shrimp tea.

In the early days of Birkenhead there was no bank and as late as 1850 the only one was that of the North and South Wales Bank in Hamilton Square, and so one day when Mr Brassey decided that he needed a bank account he travelled to Chester and walked into the Chester branch of Dixon & Wardle's Bank, Northgate Street, and asked to see the Manager. When he was interviewing the well mannered, soft spoken, young man did the manager, one wonders, realise that the young fellow sitting opposite him would within a few years command the largest industrial army in the world and that he would be responsible for more gold flowing through the bank in a year than passed through many European treasuries and yet he was so loved by his navvies that when he lay on his deathbed many of them travelled great distances to see him and to weep?

T. BRASSEY'S FIRST RAILWAY CONTRACT

In 1827 George Stephenson was building the world's first commercial passenger railway between Liverpool and Manchester. The Liverpool terminus has long since gone but the Manchester Station, Liverpool Road, still stands, forlorne and neglected, the oldest passenger railway station in Europe left to rot by British Railways. Only a nationalised undertaking could be so uncaring abut part of our heritage.*

Stephenson was building the Sankey Viaduct and went to see Brassey in Birkenhead in search of load bearing stone. He had been told of the hard Kyper sandstone being cut in Storeton Quarries and Brassey was looking after these quarries. George Stephenson was impressed by Thomas Brassey and so he suggested that Brassey should try his hand at railway building and so fired his imagination that Brassey started preparations and made a bid for the contract to build the Ditton Viaduct on the Grand Junction Railway.

The Ditton tender was unsuccessful but in 1834 Brassey was invited by Joseph Locke to tender for the Penkridge Viaduct between Stafford and Wolverhampton on the Grand Junction Railway. Brassey put in a price of £26,000 but Locke thought this was too high so he visited Brassey in Birkenhead and went over an ammended specification which reduced Brassey's price to just over £20,000. On the strength of this the contract was awarded to Thomas Brassey and he was in the railway building business.

The Penkridge Viaduct still stands a superb example of Brassey's workmanship and it was the beginning of a very fruitful partnership between Locke and Brassey. Brassey once said that what he built was not intended to last ninety-nine years, but after that period it should be as good as when it was built. There can scarcely be a reader of these notes who has not travelled on a Brassey built railway line and almost all of his work still stands after one hundred and forty years apart from those lines closed by British Rail.

*1988 Note. Happily since this was first written in 1976 British Rail have sold the station to a trust who have restored the station to its original splendour.

In 1835 Locke again invited Brassey to tender for 36 miles of the London and Southampton Railway from Basingstoke to Winchester and again Brassey was successful. The cuttings on this line are amongst the most immense of any railway, but again the contract was carried out to the entire satisfaction of Joseph Locke and more and more he became impressed with Brassey's vigor, energy, ability, honesty and integrity.

Thomas Brassey's energy was fantastic, at the peak of his career he employed 85,000 men all over the world and yet he did not have an office organisation as such. He did not employ a secretary and wrote his own correspondence by hand. Wherever he went he carried a bag containing letter and writing material and all letters were answered immediately. At one time he shared a shoot in Scotland with Joseph Locke and one day while they were out shooting there was a lull in the sport and after a short time Locke missed Brassey and then found him sitting on an old box in a delapidated shepherd's hut quietly writing letters. A Dr Burnell who travelled with him in Italy recalled that one evening they had dinner and at 9 o'clock Dr Burnell went to bed and when he came down in the morning found thirty one letters in Brassey's handwriting to agents and sub-contractors all over the world.

Henry Harrison, Thomas Brassey's brother-in-law recorded that "I have known him come direct from France to Rugby, having left Havre the night before, he would have been engaged in the office in London all day; he would then come down to Rugby by the midnight mail train and it was his common practice to be on the works by six o'clock the next morning. He would frequently walk from Rugby to Nuneaton in the afternoon, a distance of sixteen miles. Having arrived at Nuneaton he would attend to his work and the same night would go to Tamworth by road. The next morning he would be up early and would walk the line they were building from Tamworth to Stafford; that evening he would take the train from Stafford to Lancaster to inspect the line from Lancaster to Carlisle which he was building. This route was one he frequently followed."

The railway contracts followed thick and fast and in 1841 he started the first railway that he built abroad. He was awarded the 82 mile Paris to Rouen railroad, the first in France, this is a story in itself, but sufficient to say that by 1848, seven years later Thomas Brassey had built three quarters of the entire French railway system.

Brassey built thousands of miles of railways in Britain, France, Spain, Norway, Austria, Germany, Denmark, Italy, India, Mauritius, Australia, Canada, Poland, Russia and South America. All his work was to a very high standard of workmanship and if he made a mistake he put it right at his own expense; his reputation for honesty and integrity spread all over the world and he became a national figure.

CANADA WORKS, BIRKENHEAD

In 1852 Thomas Brassey teamed up with Sir Morton Peto and Edward Betts. Peto's brother in law and formed Peto, Brassey and Betts. They were awarded nine contracts that year one of which was for the construction of the 540 mile Grand Trunk Railway of Canada. This line was to be the longest railway to be built at that time and was to run from Quebec to Lake Huron through virgin forests and across the mighty one and three quarter mile St Lawrence River at Montreal.

Thomas Brassey decided that they would have to build a new works to make the vast amount of railway equipment needed for this enormous undertaking and he decided to build his works in Birkenhead because the Mersey was excellent for shipping goods out to Canada; it was in an area where there was excellent labour—still true—and he knew it well because it was his, and his wife's, old home town.

The way Thomas Brassey and his men set about building a large and sophisticated engineering works in a matter of months is an object lesson for modern industry.

Brassey recalled his brother-in-law George Harrison who was working for him in Canada and told him what he wanted and put him in charge of the new works which was to be called Canada Works in honour of their biggest contract.

George Harrison returned to England late in May 1853 and on the 29th May he toured the

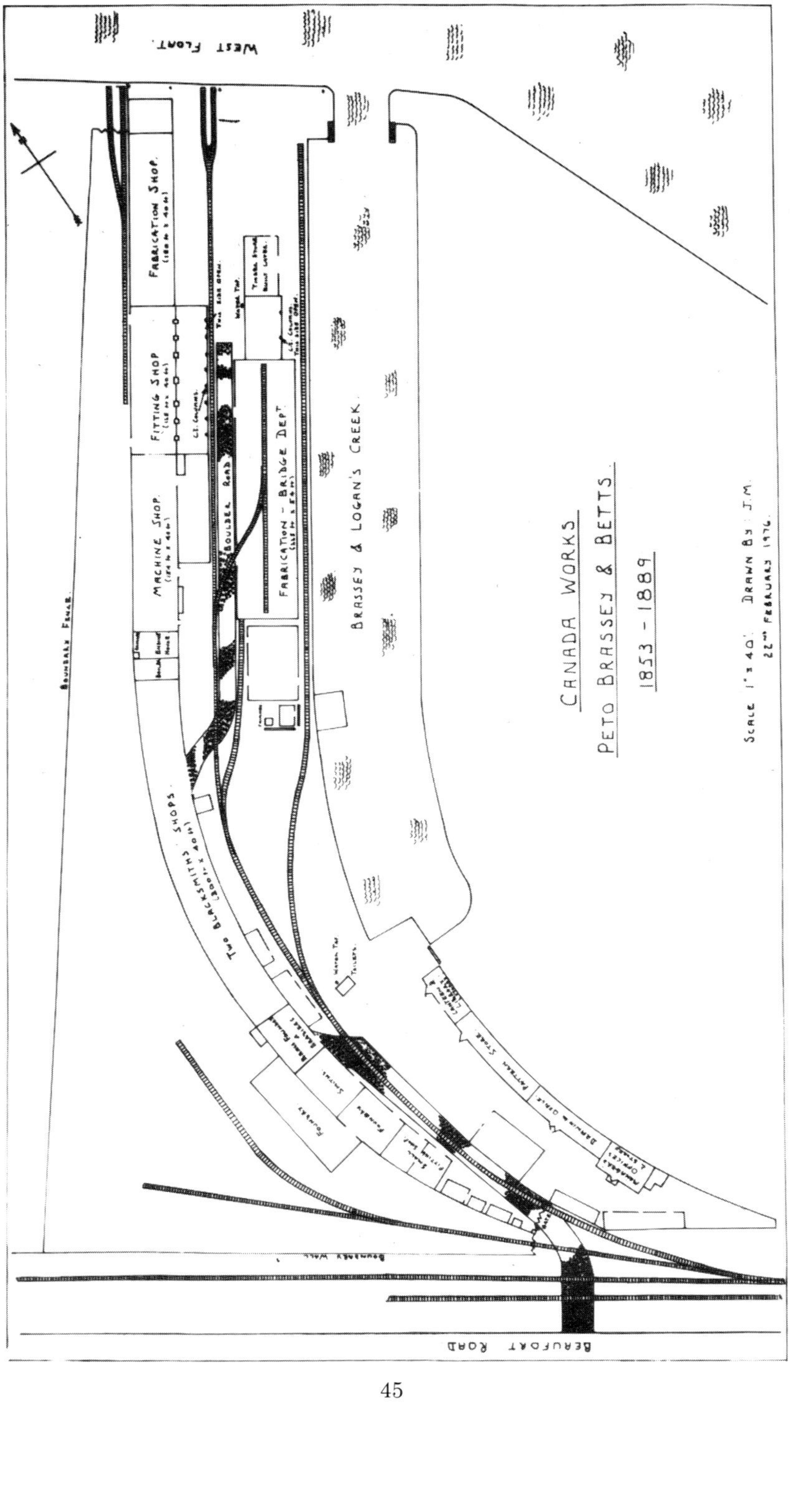
CANADA WORKS
PETO BRASSEY & BETTS
1853 - 1889
WEST FLOAT
BRASSEY & LOGAN'S CREEK
FABRICATION SHOP
FITTING SHOP
MACHINE SHOP
BOULDER ROAD
FABRICATION - BRIDGE DEPT
TWO BLACKSMITHS' SHOPS
BOUNDARY FENCE
BOUNDARY WALL
BEAUFORT ROAD

dock area of Birkenhead with a building contractor, Mr Meakin, looking for a suitable piece of land. Late in the morning they found a piece of land which, although not ideal in all respects, would meet their needs. It was excellent in that it fronted on the Great Float and sea-going vessels could moor against the quay and that there was another dock running the length of the land, but it was banana shaped which is not ideal for an engineering works, but it was the best they had seen. The land covered about nine acres and although there was no access road there was a railway line nearby, built by Brassey in 1849.

In the afternoon Mr Harrison spent one and a half hours doing sketch drawings for the new works and was given a budget price by Mr Meakin. That evening he travelled to London by train and saw Thomas Brassey early the next morning, they went over the plans and he returned to Liverpool the same afternoon by the express train.

The next day George Harrison bought the land and placed a verbal order with Mr Meakin, who started work immediately and the first brick was laid on the 4th June 1853, exactly a week after they had first found the land. Work was pressed on and the new factory was virtually finished by October of the same year. The whole project from finding the land to taking over, occupying only 5 months. It would take longer than that now to obtain planning permission.

The new works were an absolute model of what an engineering works of that era should be, the 13″ walls were made of red pressed brick, the roof was lined slates, the floors were flagged and illumination by day was by generous windows and rooflights, and at night by gas lights.

The main shop was 900′ long and 40′ wide and the works comprised iron and brass foundries, blacksmiths shops, copper smiths, machine shops, fitting out bay, fabrication shop, woodwork and pattern shop, engine house, stores and drawing office. The very latest machines were bought and because, at that time, American woodworking machines were in advance of ours—because the shortage of skilled labour in America had encouraged the development of machinery—Brassey sent two men to America to buy the latest machines and they brought back morticing and planing machines.

The machine shop was equipped with a number of Whitworth lathes, including one which could turn a wheel 7′ diameter without any adjustment of the bed. Planing machines, slotters, drills, etc, were all driven by two 40-hp steam engines.

The fabrication shop where the bridges were made and boiler work was done and which was run by William Evans and William Heap, was equipped with three furnaces for heating rivets, one furnace for angle and "T" iron, eight punching and shearing machines, two sets of bending and plate rollers, two rivet making machines, two automatic rivetting machines and power was provided by a 35-hp steam engine. Later the number of machines was doubled. One of the presses was the immense Roberts Jacquard press already referred to and this probably had a separate drive. The great problem with this machine as already mentioned, was to find enough long runs to keep it fully occupied, but on the Victoria Bridge, the Jumna and other great bridges it was ideal and its accuracy in punching holes is remarked on by James Hodges, the Resident Engineer on the Victoria Bridge in the extract from his report which we print later.

The fitting shop for locomotive production was 150′ long and comprised ten engine building pits and was designed to produce forty locomotives a year; this capacity was to be fully employed over the next few years because the Canadian railway alone needed three hundred engines over the next eight years.

There were two well equipped blacksmiths' shops. The main shop was equipped with forty furnaces, each with its own anvils, blacksmith's tools and all were fully manned; there were also two small, but efficient, steam hammers. The smaller smithy was quite interesting because that was where the wrought iron locomotive wheels were forged. The wheels were made by heating up individual bars and then welding together two bars to form a "V"; then another "V" was made and then welded to the first and so on until all the spokes were welded by the smith into position. At the end of each spoke was a knob of metal and these were welded together and then the rim was forged. The rough forging was then taken to the machine shop and the wheel was turned up by one of the many wheel lathes.

Down the centre of the works was a cobbled road made up of rounded boulders, probably brought to Liverpool by sailing ships that left these shores loaded with manufactured goods

Excavating the cobbled road at Canada Works 1971.

Relaying the boulder road from Canada Works at Britannia House, Hoylake. The water fountain stood for over a century in Pall Mall, Liverpool.

and sailed back with the stones as ballast. These stones were beautifully laid and when in 1971 we bought a stretch of the road from Mersey Docks & Harbour Board, so we could relay them in the garden in front of our offices at Hoylake, we had great difficulty in lifting them, because although they were rounded, they had been cleverly interlocked. In the end we were making such little headway with picks and crowbars that we brought in a bulldozer, but even so they did not give up easily.

On the right of the works was the drawing office and stores and adjoining them two houses and in one of them lived Andrew Alexander M.A., who was a sub-manager with William Heap,

*The site of Canada Works from the West Float. Brassey and Logan's creek is on the left. Today only the quay and two buildings remain of that remarkable engineering works.**

*1988 Note. Since 1976 Logan's Creek has been filled in.

under the Manager, George Harrison. Later William Heap was made Manager of the bridge department which was a most important section of the works.

Canada Works was well in advance of its time because they also had a test and inspection department where the plate, bar, rod and other raw materials were tested for tensile strength, impact and compression.

BRASSEY'S TREATMENT OF HIS MEN

When Canada Works opened in October 1853 there were about four hundred men in the machine shop and one hundred and twenty-three men under William Heap in the bridge department but later these numbers were increased to about one thousand two hundred men, evenly split between the two sections. The average wage paid was £1 a week and considering that there were a considerable number of young apprentices the skilled men were taking home much more than a pound a week which was a very good wage for the day. There was little wrong with the food that they ate because as previously mentioned Brassey's navvies ate two pounds of beef a day for a start.

Thomas Brassey and his Managers differed from some of the other prominent engineers of the time, such as George and Robert Stephenson and Isambard Kingdom Brunel, in their attitude towards the education of their workmen and they were not alone in this respect because many employers in Victorian times provided their own schools or subsidised British Schools. In Turton, Lancashire as early as 1825 it was reported that five hundred and twenty-five out of five hundred and thirty-two employees could read and that there was a library at the mill.

The Stephensons however and Brunel (even great men have some failings) did not think it necessary for mechanics to be educated beyond their immediate work, but Thomas Brassey, George Harrison, William Heap and Andrew Alexander took the opposite view and at the very outset they built at Canada Works a canteen and a library for the workmen.

The Management gave £75 to start the library and provided 600 well selected volumes, the men gave a concert and ball in one of the large halls of Birkenhead for five hundred of the men and their wives and they raised a further £12. On the 20th October 1853, the reading room was opened and the men found available a wide selection of books as well as a selection of local and national newspapers. In those days advertising revenue had not reduced the price of a newspaper to the fantastic bargain of today and for example the "Liverpool Courier" cost 5d which in those days was a tremendous sum.

The men paid a penny a week towards the reading room, which was open every meal time and was run by a committee of workmen. Books were given out twice a week and at the same time those that had been borrowed could be returned. It is of interest to note that with the exception of the Waverley Novels, most of the books borrowed were of an educational nature, particularly those dealing with engineering and mechanics.

"THE LADY ELGIN"

The speed of Canada Works was breathtaking to a modern observer because the first locomotive was tested in May 1854 exactly a year after George Harrison and Mr Meakin walked on to the barren, banana shaped field which they transformed into Canada Works.

Engine No. 1 was called "The Lady Elgin" after the wife of the Governor General of Canada and it passed its tests with flying colours, every part of the engine except the tubes and plates had been made in Canada Works and that included the wheels and even the rivets.

"The Lady Elgin" was a bogy chassis passenger locomotive built for the "Grand Trunk" and if one is permitted to talk about such intimate details of a lady, her vital statistics were:

Cylinder	15″ bore
Piston Stroke	20″
Tube Size	$1\frac{1}{4}$″ diameter

No. Tubes	175 in the smoke box
Total Heating Surface	948 square feet (872 sq. ft. in the smoke box + 76 sq. ft. in fire box)
Wheel Gauge	5′–6″
Leading Wheels	3′–6″ diameter
Trailing Wheels	6′–0″ diameter
Load	22/23 passenger per carriage
Speed	40 m.p.h.
Fuel	Wood burning

After the highly successful trials on the specially laid 5′–6″ track, "The Lady Elgin" gave a mighty shriek of triumph on her whistle and she was decorated with oak branches by the workmen and there was loud cheering. That Monday evening George Harrison took Andrew Alexander, William Evans, William Heap and the Foremen to a dinner at the Woodside Hotel, Birkenhead and the following Saturday they gave a celebration dinner for all the men at an hotel in Birkenhead.

At the same time, engine Number 2, "The Lord Elgin", was almost finished and there were five other passenger and five goods locos in the course of erection. The goods engines differed from the passenger engines, only in the cylinder bore, which was 16″ and the driving wheels, which were 5′ in diameter.

By the end of 1854, ten locomotives had been shipped to Canada, eight more goods engines were ready for shipment, there were sixteen others being made and No. 35 was being started.

"The Lady Elgin". The first engine to be built by Canada Works, successfully completed her trials in May 1854 exactly a year after the site for Canada Works was found.

Constructing the Victoria Bridge, Montreal 1853–1858.

In the bridge department William Evans and William Heap had not been idle, because they had been given several jobs to do and one of them was the construction of the Victoria Bridge, which was to be almost two miles long—by far the longest bridge in the world and even now, one hundred and twenty-five years later, is in the top five longest bridges ever built and will stand as an all time record as the longest tubular bridge ever built.

The site problems were immense as we shall see later, but William's task was gigantic because over three million holes had to be punched and drilled in hundreds of thousands of components and shipped to a site thousands of miles away in such a way that they arrived on site on time, in the correct order and somehow they had to be readily identified by the men working on the bridge.

Mr Heap's first task was to set up a system for marking and tagging the hundreds of thousands of plates and girders, each piece was laid out on the floor together with its companion pieces and then properly marked so that it could be located quickly on the other side of the Atlantic.

Mechanical digger built at Canada Works, Birkenhead, working on the foundations for the Victoria Bridge 1853.

Victoria Bridge 1853–1858.

The plates were ordered from Staffordshire and when they arrived they were then rolled flat and wherever possible they were punched on the Roberts Jacquard press, which had been programmed for this immense task. Wherever possible they built up small assemblies at Canada Works to save site work.

By the time "The Lady Elgin" had been tested in May 1854, William Heap had already shipped out one span of 155′ and they were busily engaged on the next span. But it should be remembered that this was not the only work they were doing because they had made and shipped by December 1854 a 120-hp winding engine and a large boiler for a works in Derbyshire, as well as working on a number of other smaller bridges for the Grand Trunk.

CONSTRUCTION ON SITE

William Heap visited Canada during the construction of the Victoria Bridge and saw at first hand the tremendous difficulties that had to be overcome by Brassey's engineers and navvies, he was also able to see that the system he had devised was working splendidly.

The winter ice in the St. Lawrence poses a threat to any bridge because it piles up to a height of twenty feet or more and in the spring with the thaw, these tremendous weights of ice are carried forward by the fast current and form a ferocious battering ram. Stephenson designed his masonry piers so that they are of the most robust construction and facing upstream; the piers are wedge-shaped to cut the ice and guide it around the piers.

Tremendous problems were encountered in building the piers because before construction

Emigrants' Stone, Montreal.

was started it was thought that the bottom of the river was rocky and firm, but in many places they found it was comprised of loose boulders, some of immense size and these had to be cleared away before the coffer dams could start. Some idea of the difficulties can be gained from a report in the "Engineer" dated 9th September 1859 in which it is mentioned that one boulder alone weighed 24·5 tons and this had to be moved from the bottom of an icy, fast flowing, river.

The workmen used one of the boulders they had taken from the bottom of the river as a memorial stone to the 6,000 emigrants who lost their lives of "ship fever" or typhoid in 1847. on the voyage to the New World, and they mounted this on a plinth near to the railway. In the early days it was known as "The Emigrants' Stone", but now is known as the "Irish Emigrants' Stone".

Before the piers could be made, coffer dams had to be constructed in the river and great difficulties were experienced in their building, because even in summer there were dangers such as the occasion when logs swept down the river destroyed the coffer dam made for pier 13. Once the dams had been constructed around the area where the piers were to be built, then the water had to be pumped out and "The Engineer" of 18th March 1859 reported that the pump used was: a portable centrifugal pump, light enough for a man to carry over his shoulder. Astonishingly enough this pump had a capacity of 800 g.p.m. and "could pass stones up to 6″ square". This pump lowered the levels of the coffer dam at the rate of 2′ an hour and in 10 hours the first coffer dam was pumped dry. Regrettably "The Engineer" did not give the manufacturer of this pump, otherwise we should now be writing for the Agency.

The Victorian Engineers needed little excuse to give a party; on reflection times have not

Log jamb damages Coffer Dam on Pier 13.

changed all that much; but they were unusual in the way that they linked the celebration with an engineering achievement. After the heartbreaking struggle by Sir Marc Brunel to build the Thames Tunnel, which was eventually finished by his son, Isambard Kingdom Brunel (we are sorry to keep on repeating his full name, but it sounds like a roll of thunder worthy of the Victorian engineers), they gave a magnificent dinner in the tunnel itself, with an enormous table laid with a white cloth, gleaming silver and decanters.

When Thomas Brassey finished the Paris to Rouen railway in May 1843, there was a ceremonial opening of the line by a Bishop, but Brassey gave his men an open air banquet where 600 men sat down. A whole ox was roasted by three French chefs but one of Brassey's men turned the spit and in a contemporary drawing he is shown drinking wine from a bottle. The French became so agitated at the prospect of six hundred British navvies going on the rampage that they ringed the field with their cavalry.

The men of Peto, Brassey and Betts were not to be outdone because they gave a party at the bottom of one of the pumped out coffer dams, and as "The Engineer" commented at the time, "It was an unusual setting, with the mighty St Lawrence swirling outside to a height of thirty feet."

It is a pity that this direct linking of an engineering achievement with a celebration has gone out of fashion, because one can think of dozens of good excuses for a banquet for those who have achieved much in recent times that has passed by with nothing to mark the event. For example when man first landed on the moon—incidentally an engineering rather than a scientific achievement—would it not have been glorious if the Astronaut had climbed down the steps and set out a table for a celebration dinner, instead of mumbling those embarrassing words dreamed up by some hack in the State Department months earlier about "A small step for man . . .".

Please do not write to us as we do appreciate the problems, but engineers are not too bad at solving them, particularly if they think that a party is at stake!

Having overcome the problems of the coffer dams James Hodges the Resident Engineer for Peto, Brassey and Betts, then had to find the stone to build the piers and after a long search he found a hard limestone about sixteen miles upstream of the bridge, but it was on land owned by the Red Indians.

Mr Hodges arranged a meeting with the thirteen chiefs involved and disappointingly enough for the writer, he found that the most convenient time to meet them all together was on Sunday morning after church! If you wonder why we find this remarkable remember that this was twenty-three years before General Custer made his last stand before the Sioux Indians, but then the British have always been better Colonisers than the Americans, or the Russians come to that.

Instead of the proud befeathered Chiefs of the writer's most loved films, the Red Indians turned out to be miserable, dirty looking old men, smoking short clay pipes, no doubt made in England. On their side, the Chiefs were not at all impressed with James Hodges; they complained that he was far too young to negotiate with them—"Why, he has not even grown whiskers!" Mr Hodges gravely assured them that he was over forty years of age. Thus assured the Chiefs then agreed to negotiate and eventually agreement was reached and after the Red Indians had been paid, the stone was cut and transported to the site of the bridge.

The volume of stone cut for the piers of the Victoria Bridge was 2,713,000 cubic feet, or

Invitation to the Opening of the Victoria Bridge.

enough to cover a six acre field to a height of eleven feet, and all this was done without our modern means of transport.

After the piers were built, work started on the first spans and Brassey's men built temporary wooden lattice bridges, supported in winter by the frozen river, and on this they built up the world's biggest "Mechano Set" sent out with full instructions by William Heap in faraway Birkenhead. To facilitate the assembly of the parts they had a number of quite advanced mechanical devices built for them by their colleagues on the Great Float, one of these was a travelling crane which ran on tracks run on the temporary wooden structure and the plate layers and rivetters could move this by operating two cranked operating handles. They also had the advantage of the first mechanical shovel ever made by a British Contractor, this was improved on site and the men found it was excellent for digging out very hard material, but in light sand and gravel it was easier, and cheaper, to use manual labour.

These grand Victorians were building the longest bridge in the world and as well as overcoming the engineering problems they had to combat a climate which in winter was to say the least, inhospitable. The temperature dropped so low that if a man touched the steel with his bare hands the flesh stuck to the frozen metal; icy winds caused frost bite and they suffered from dysentry and cholera. At one time over a third of the men working on the bridge were sick with cholera and many died, but Brassey's men had guts and they built their bridge well and on time.

Great Tubular Bridge at Montreal, Canada.

DEC. 20, 1856.] FRANK LESLIE'S ILLUSTRATED NEWSPAPER. 41

GREAT TUBULAR BRIDGE, GRAND TRUNK RAILWAY, MONTREAL, CANADA.

GREAT TUBULAR BRIDGE AT MONTREAL, CANADA.

THE intimate connection between Canada and the United States, since the reciprocity treaty, makes any advancement in internal improvements on the part of Canada of deep interest to the people of the United States. No one can really calculate the happy results North to Georgian Bay, ninety-six miles; the Montreal and Bytown railway, 130 miles in length; the Prescott and Ottawa railway, fifty miles long; the Cobourg and Peterborough railway; the Port Hope and Lindsay railroad, thirty-six miles in length; the Buffalo, Brantford and Goderich railroad, connecting Buffalo with Lake Huron, 160 miles long; the Brockville and Ottawa, a feeder of the Grand these roads. A line is also in course of construction from Halifa[x] in Nova Scotia, to connect with the Grand Trunk which is designe[d] to obviate the difficulties arising from the long-ice blockade of th[e] St Lawrence.

Report by James Hodges

James Hodges wrote a report of the building of the "Great Victoria Bridge" and we give the following extract:

CONSTRUCTION
OF THE
GREAT
VICTORIA BRIDGE
IN
CANADA.
BY

JAMES HODGES,

ENGINEER
TO

MESSRS. PETO, BRASSEY, AND BETTS,
Contractors.

London
JOHN WEALE, 59, High Holborn.
1860

The whole of the iron-work for the tubes was prepared at the Canada Works, Birkenhead, where a plan or map of each tube was made, upon which was shown every plate, T bar, angle iron, keelson, and cover plate, in the tube, the position of each being stamped and marked upon it by a distinctive figure, letter, or character. As the work progressed at Birkenhead, every piece of iron, as it was punched and finished for shipment, was stamped with the identical mark corresponding with that on the plan; so that, when being erected in Canada, although each tube was composed of 4,926 pieces, or 9,852 for a pair, the workmen, being provided with a plan of the work, were enabled to lay down piece by piece with unerring certainty till the tube was complete. To an uninitiated spectator this proceeding would appear as complicated and hopeless a task as the fitting together of a Chinese puzzle; but to such perfection did they arrive at Birkenhead in making the plans, in preparing and punching the iron, and in shipping it, that when it arrived in Canada (where the iron for each tube was, as it arrived, sorted and stacked separately for use), the workman being provided with the plan would proceed with his work throughout, and never put a piece in the wrong place, nor have to alter a single plate. It was not uninteresting to watch the gradual diminution of the pile of iron on the platform as the work progressed, and eventually to see the last piece taken to fill up some out-of-the-way hole or corner, and then to hear for certain that the tube was completed.

As soon as it became evident that the centre piers, 12 and 13, would be finished in time to receive the centre tube if erected during winter, two temporary piers were commenced in the centre opening, the intention being to carry up these erections in cribwork till they were high enough to support a continuous Howe-truss extending across the opening, the ends of which were to rest upon large stone corbels left for that purpose on the face of the piers.

At this period the sectional area of the river was so much diminished by the temporary works now in the deep water, that the current was increased greatly, and it was almost impossible to hold a crib of any dimensions in such a current till it was sunk in position. The plan pursued in this instance was to build the crib in the eddy on the south side of No. 12 (the stream which set partially across the piers being favourable to such an operation), and then to strut it off into the current from the dam, and sink it as speedily as possible. Of course heavy moorings were laid out ahead to hold the crib in place against the stream, and chains were likewise made fast to the bow.

I have before noticed that the whole of the iron work for the tubes was prepared at the Canada Works, Birkenhead (an establishment erected by Messrs, Peto, Brassey, and Betts expressly for the manufacture of the bridge work and rolling stock for their Canadian contracts). At these works every plate, &c., was finished ready for putting in place. I have likewise endeavoured to show with what care and accuracy the whole of this was done; but I trust I may be excused for again drawing attention to the extraordinary perfection attained in the preparation of this ironwork. In the centre tube, consisting of 10,309 pieces, in which were punched nearly half a million of holes, not one piece required alteration, neither was there a hole punched wrong! The importance of this accuracy may be estimated by considering that had any portion been carelessly prepared, or even wrongly marked, a failure might have been the result, involving the delay of a year in the opening of the bridge, and a loss of many thousands of pounds. Therefore, to Mr. George Harrison, the manager of the Birkenhead works, &c., and to his able assistants, Messrs. Alexander and Heap, is due as great credit for the successful completion of this work as to those engaged in its erection. For details of tubes, see Plates 24 and 24a.

The bridge was first opened for the passage of trains on the 19th of December, 1859. The formal inauguration by his Royal Highness the Prince of Wales, who visits Canada for the purpose, is appointed for the 25th of August, 1860, when, under God's blessing, this work, of such great social and international importance, will be duly dedicated to the great purposes for which it is designed.

The following are the inscriptions at the entrance of the Bridge:—

[ON THE OUTER LINTEL.]
ERECTED A. D. MDCCCLIX.
ROBERT STEPHENSON AND ALEX. M. ROSS,
ENGINEERS.
[ON THE INTERIOR LINTEL.]
BUILT
BY
JAMES HODGES,
FOR
SIR S. MORTON PETO, BART., THOMAS BRASSEY,
AND
EDWARD LADD BETTS,
CONTRACTORS.

DATES.

First part of north abutment coffer-dam towed into place 24th May, 1854.

First stone of bridge laid 20th July, 1854.

First train passed over the bridge 17th December, 1859.

DIMENSIONS, WEIGHTS, &c.

Total length of tubes, 6592 feet.

Total length of bridge, 9144 feet.

Height of bottom of centre tube above surface of water, 60 feet.

Height of bottom of tubes at abutment, 36 feet.

Rise of tubes to centre, 1 in 130.

Weight of iron in tubes, 9044 tons.

Number of rivets in tubes, 1,540,000.

Painting—number of coats, 4; number of acres in each coat, 32. Total acres, 128.

Number of piers, 24.

Number of spans, 25; twenty-four from 242 to 247 feet each; one, 330 feet.

Quantity of masonry in piers and abutments, 2,713,095 cubic feet.

Greatest depth of water, 22 feet.

Average rate of current, seven miles an hour.

Quantity of timber in temporary works, 2,280,000 cubic feet.

Quantity of clay puddle used in dams, 146,000 cubic yards.

FORCE EMPLOYED IN CONSTRUCTION.

Number of steam-boats, 6.
Number of barges, 75.
Tonnage of ditto, 12,000.
Power of steamers, 450 H.P.
Number of men employed, 3040.
Number of horses, 144.
Locomotive engines, 4.

The Grand Trunk Railway was not well conceived by the promoters and it was one of the most difficult contracts that Thomas Brassey ever tackled. Financially the building of the Victoria Bridge nearly broke Brassey, and incidentally, the backers of the railway, but as a piece of engineering it was a masterpiece and as such it was widely acclaimed.

"The Engineer" in an editorial to mark the opening described it as "The greatest example of engineering construction in the world". When the Prince of Wales opened the bridge on the 19th December 1859 he said of Peto, Brassey & Betts of Birkenhead, "They left behind an imperishable monument of British skill, pluck, science and perseverance".

The final word must be given to Robert Stephenson who at a dinner given for him later in Toronto by the Engineering Profession of Canada said of Peto, Brassey and Betts' work on the Victoria Bridge "The Contractors left even the Engineers themselves little more than the poetry of engineering".

The original Victoria Bridge gave good service for over forty years and was only converted in 1892 to a lattice girder bridge on the original piers because the original specification called for a single line track and this was later converted to a double track with a subsequent speeding up of traffic, and also the two mile line tubular bridge caused smoke problems in the days of steam.

The links that William Heap forged in Canada whilst working on the Victoria Bridge remain with us to this day.

CANADA WORKS AND THE CRIMEAN WAR

The Crimean War has stayed alive in public memory largely because of Alfred, Lord Tennyson's poem "The Charge of the Light Brigade" and for the work done by Florence Nightingale in setting up hospitals to nurse the sick and the wounded; but few know that the British Army might well have been defeated but for the organising ability of Canada Works and the stupendous efforts of Thomas Brassey's engineers and navvies. At the time Brassey and his men were national heroes but now their deeds are almost forgotten.

In 1854 a British Army of 30,000 men, together with a French Army, were fighting the Russian Army in the Crimea and at the end of the year were beseiging Sevastapol. All the supplies for the army had to be dragged, sometimes axle deep in mud, over a single road from Balaclava, the nearest part in British hands.

Unlike our modern army, the British Army of 1854 was badly led, ill clad and equipped and the casualty lists grew alarmingly as men died of wounds, frostbite, cholera and starvation. It was plain to all that the British Army was about to be defeated by the same factors that had beaten Napoleon earlier, and Hitler later—the Russian winter and overstretched supply lines.

The British Army had not fought a major battle since Waterloo, thirty-nine years before and although its Officers were brave enough they had become incompetent. Cecil Woodham Smith summed up the officers who fought in the Crimea as follows: "They affected elegant boredom, yawned a great deal, spoke a jargon of their own, pronouncing 'r' as 'w', saying 'vewwy', 'howwid' and 'sowwy' and interlarded their sentences with loud and meaningless exclamations of 'Haw, Haw!'. Their sweeping whiskers, languid voices, tiny waists, laced in by corsets and their large cigars were irresistable; frantically admired and as frantically envied. Magnificently mounted, horses were their passion, they rode like the devil himself, and their confidence in their ability to defeat any enemy single handed was complete. Cavalry officers were saying in London drawing rooms that to take infantry on such a campaign was superfluous; the infantry would merely be a drag on them and had better be left at home."

A French Commander after praising British horses and their riders said: "The British cavalry officer seems to be impressed by the conviction that he can dash or ride over everything as if the art of war were precisely the same as that of fox hunting." How right he was.

We can question their competence but not their bravery. Lord Cardigan who commanded the Light Brigade during the famous charge knew it was madness to charge into the "valley of death", but after receiving the mistaken order he saluted and said to himself "Well here goes the last of the Brundenells"—his family name; prepared to charge in perfect drill book order "The Brigade will advance, walk, march, trot."

Beautifully dressed, stiffly upright, never turning his head, Lord Cardigan charged down the mile and a half of murderous Russian shot and shell right to the Russian guns. Out of seven hundred horsemen only one hundred and ninety-five returned and five hundred horses were killed. Miraculously, Lord Cardigan was not hit and charged right through the astonished Russian gunners.

After returning to his own lines he rode back to his private yacht moored in the harbour, had a bath and drank a bottle of champagne.

The immortal words of the French General who watched the charge from the heights summed up the waging of the Crimean War 'C'est Magnifique, mais ce'nes pas la Guerre", it was magnificent but it was not war and the public were stirred into action by the reports of William Howard Russell to "The Times" and the photographs taken by one of the earliest war photographers Robert Fenton.

The value of a free press to a Nation was clearly illustrated by Mr Russell, because he exposed the criminal incompetence of Government and Army and people in many walks of life started to take action.

One of the many difficulties encountered by the British Army was that our troops still had to fend for themselves, whereas in the French Army one soldier cooked for twelve. Our methods worked well enough in kinder climates, but not in the depths of a Russian winter and the outcry in the Press prompted Alexis Soyer the chief cook at the Reform Club in London to volunteer to go to Scutari and there he devised the field kitchen and new methods of cooking for an army. Florence Nightingale went out to organise Field Hospitals, and incidentally to start out on a path which was to revolutionise nursing. Another group of men who thought they could help were Thomas Brassey, Morton Peto M.P., and his brother-in-law Edward Betts who were trading as Peto, Brassey and Betts of Canada Works.

On the 30th November, 1854 Morton Peto told the Government that Peto, Brassey and Betts could overcome the supply problem by building a thirty-nine mile long railway from Balaclava to all parts of the front and to cut delay to the minimum they would do this at no profit, would provide men and materials and would run the line when it was finished. News had reached the Government a little earlier that the plight of the Army was desperate. On 14th November there had been a dreadful blizzard; the troops before Sevastapol only had thin clothing, no tents and no fuel. The icy winds had blown the blankets off the wounded and many froze to death, there was little food for the men or fodder for the horses and soon there were 8,000 men in hospital and replacements from England often died before they reached the front.

Although many doubted if it would be possible to build a railway quickly enough to save the situation in the depth of a Russian winter, the Government were absolutely desperate and they gratefully accepted the offer. Thomas Brassey then set his superb organisation to work and what a difference between the criminal incompetence of the Government employees and those of the Birkenhead firm.

On the 30th November, 1854 when Brassey's offer was made the rails and sleepers didn't exist, but as soon as Brassey gave the order his men spent the first day drawing up a list of the materials they would need and where they could be produced. To save time they chartered, or bought, vessels near to the location of the materials they were to carry. For example some of the rails were to be produced at Walker on Tyne and so they bought the screw steamer "Hesperus" which was just being completed by Messers Mitchell in the next yard on the Tyne and the rails were loaded warm from the furnances into the hold of the "Hesperus".

In three days Peto, Brassey and Betts had bought, or chartered, a fleet of ships, all on verbal order and without any security. All the shipowners and shipyards worked night and day to convert the ships to their new use. and cabins and bunks for Brassey's engineers and navvies were built in a few days. The first ship bought by Canada Works was the clipper "Wildfire". She was bought from Tonge, Curry & Co. of Liverpool for £4,500 and was sailed over to the Great Float, Birkenhead for conversion and loading. Needless to say, she was the first vessel ready to sail, but she was closely followed by others and the fleet comprised:

"Wildfire"	Clipper sailing ship	457 tons	
"Mohawk"	Clipper sailing ship	800 tons	
"Lady Alice Lamberton"	Screw steamer	511 tons	90-hp
"Great Northern"	Screw steamer	578 tons	90-hp
"Earl of Durham"	Screw steamer	554 tons	90-hp
"Baron von Humboldt"	Screw steamer	420 tons	60-hp
"Hesperus"	Screw steamer	800 tons	150-hp
"Prince of Wales"	Paddle steamer	627 tons	120-hp
"Levant"	Paddle steamer	694 tons	500-hp

The "Prince of Wales" was the same vessel that had suffered the fortuitous accident at Millwall that had given Robert Stevenson the idea of the tubular bridges at Conway and Menai.

In three days verbal orders were placed for vast quantities of material and where there was not sufficient time to make the large items of plant or rolling stock they were found and bought. For example the steam machinery was at work on the Victoria Docks, London, which Brassey had built, and as it was needed for the Crimea it was bought.

The loading of the ships was so organised and the material was distributed such that if one vessel was delayed, or lost, then it would not endanger the whole operation.

An initial labour force of five hundred navvies and engineers was organised, many of whom

The Clipper "Wildfire" about to sail from Birkenhead for the Crimea 21st December 1854.

had worked on the Victoria Bridge and the Grand Trunk Railway in Canada and were used to working in extremely low temperatures. On Wednesday 13th December, only four days after the order was given, the first party of fifty-four navvies left Euston Station, London, bound for Birkenhead, to embark on the "Wildfire" and by Friday 15th December she was ready to sail loaded with equipment built by Canada Works. She was unable to sail however, because of fierce gales that raged in the Mersey for the next six days.

Thomas Brassey's navvies were the elite, they were paid £2-16s a week plus clothing, food and tobacco, which was a very high wage indeed for the day. Before they sailed they all signed allotment papers making over a weekly sum to be paid to their families and this averaged £1 per man.

Unlike the Government, Harrison, Heap and Alexander saw to it that Brassey's men were properly clothed, housed and fed and each man was issued with:

1 Painted bag (waterproof)
1 Painted suit
3 Coloured cotton shirts
1 Red flannel shirt
1 White flannel shirt
1 Flannel belt
1 Pair Moleskin trousers
1 Moleskin vest lined with serge
1 Fearnought slop
1 Linsey drawers
1 Blue cravat
1 Blue worsted cravat
1 Pair leggings
1 Pair boots
1 Strap and buckle
1 Bed and pillow
1 Pair mitts
1 Rug and blanket
1 Pair of blankets
1 Woollen coat
1 Pair of long waterproof boots
1 Pair of fisherman's boots
1 Pair of grey stockings
1 Pair of strong nailed boots
2 lbs tobacco for present use
and a portable stove for every ten men

The bag was to contain his kit and three days rations, and to facilitate storage no boxes—nor lumber, were allowed. For those working in water there were one hundred pairs of hip boots available.

To house the men, they provided ten portable, weatherproof huts, which were made such that they would not easily catch fire and each hut had a stove for heating together with vast quantities of coal, coke and firewood. Each hut was large enough to accommodate forty men and after they saw how successful the huts were, the Government later provided a number for the Army.

There was a plentiful supply of food and one of the new field kitchens was provided for every ten men, these were portable, but very efficient and could boil, bake or fry, in the open air.

On Thursday 19th December, "Wildfire" was still moored by Canada Works held up by the weather and so George Harrison, William Heap and Andrew Alexander gave a dinner at Gough's Hotel, Woodside Ferry to take leave of one of their colleagues, Mr Shaw who was to be in charge of the first party.

The men who had travelled up from London were accommodated in various inns in

Birkenhead and one of them wrote to the editor of the "Liverpool Mercury" just before he sailed in "Wildfire":

"Gentlemen,

You would greatly oblige the thirty miners engaged by Messrs Brassey, Peto and Betts to proceed to the Crimea, and now staying at the Sun Inn, Birkenhead in making known the excellent treatment we have received under the management of our worthy landlord, Mr Pear, The Sun Inn, Bridge Street, Birkenhead. We have likewise to return our most hearty thanks to our employers and all our supporters for the most gentlemanly manner in which we have been treated, not forgetting Mr Harkdus.

I remain your humble servant,

Evelin Flinn.
Miner.

Birkenhead
20th December 1854.

On Thursday 21st, four days before Christmas, the winds had dropped enough for embarkation to start. Mr Shaw, Mr George Arkle, a relative of William Heap's future partner T. W. Arkle, and a party of miners and navvies left the Sun Inn and walked down Bridge Street to the docks. The pavements were crowded and every window had a crowd of spectators all cheering Brassey's men on their way.

So at 11 am on Thursday 21st December, only twelve days after the order had been given, the first of the fleet, the Clipper "Wildfire", Captain Downward in command, set sail and by the 30th December, the other ships had left, or were about to leave Birkenhead, Liverpool, Hull, Sunderland and London loaded with:

50 horses.
1,800 tons of rails and fastenings.
6,000 sleepers.
600 loads of timber for bridges etc.
3,000 tons of machinery.
Fixed engines.
Cranes.
Pile drivers.
Trucks.
Wagons.
Barrows.
Blocks.
Chainfalls.
Wire rope.
Picks.
Bars.
Capstans.
Crabs.
Plant and tools.
Sawing machines.
Forges.
Carpentors tools.
100 railway tarpaulins.
A number of Dean and Adam's revolvers.

Each ship carried a surgeon and a clerk to attend to the victualling and care for the stores. The labour force comprised:

One chief engineer.
500 navvies—eventually raised to 900.
3—Assistant Engineers.

1—Chief Agent.
3—Assistants.
1—Accountant and Clerk.
Foreman and timekeepers.
30—Miners.
1—Chief Surgeon.
4—Assistant Surgeons.
4—Nurses from London's leading hospital.
Medical stocks and stores.
Borehole sinkers.
2—Railway missionaries.
A selection of books.

Balaclava 1855. Brassey's navvies at work. Note the initials P.B. & B. on the winding drum to the right of the lithograph.

Balaclava 1855. Unloading the railway equipment. Photograph by Robert Fenton—developed in his mobile caravan/darkroom. One of the first war photographs ever taken.

Balaclava 1855. Moving supplies on the railway. Photograph by Robert Fenton.

On arrival one ship was to become a stores and hospital ship and the others to be used as expedient.

This outstanding event was not reported in the Liverpool papers but "The Illustrated London News" wrote on 30th December 1854:

"The immense resources of Peto, Brassey and Betts have enabled them with very little exertion to collect, organise and ship off in an almost incredibly short space of time this important expedition for which the order was only given on 9th December.

At that time the rails were not rolled, the sleepers were uncut and the steam machinery was performing its daily function at the Victoria Docks and several of the ships were not finished."

The revolvers that were taken raised some doubts because it was felt that they might endanger Brassey's navvies and again the "Illustrated London News" wrote:

"It has been stated that the Crimean navvies are to be armed—this is a mistake—they are two valuable and expensive to be put in the way of shot if it can be avoided.

"A few arms have been sent for special cases and a few of the candidates enquired if they might have the chance of a shot at the Russians."

The ships made calls at Gibralta and Malta and the navvies went ashore, became fighting drunk and then showed the locals what they were going to do with the Russians.

Eventually the fleet arrived off Balaclava early in February 1855 and their first task was to erect their huts to protect them from the icy, howling blizzards. This was done and in no time they had heated, weatherproof accommodation. Compare this with the thin tents the poor soldiers had to live in at Balaclava.

Balaclava February 1855. Brassey's men at work building the railway. Compare the heated huts provided by Brassey for his navvies with the thin tents of the soldiers in the depths of a Russian winter.

The men named the huts "Peto Terrace", "Brassey Terrace", "Preston Hall", "Napoleon", "Victoria", "Blackwall", "Suffolk" and "Lancashire".

The man in charge of building the line was Mr Beattie, a colleague of William Heap who had worked on the "Grand Trunk". He split the labour force into two, half working by day levelling the ground, laying the sleepers and lines, and the other half worked by night, banking up and filling in between the sleepers with stones and earth. Within the first two weeks of landing they had built their encampment, several bridges and seven miles of line!

As an example of their speed of working one correspondent of the "Illustrated London News" noted that the line had to cross a small, but very marshy stream that ran into the harbour. A piledriver was landed one evening and carried piecemeal to where it was to sink piles for the bridge. The machine was erected early the next morning and all the piles were driven, the machine removed, a stout wooden bridge was constructed and the rails were laid across the bridge and one hundred yards beyond before evening!

The same publication carried a story of a "wild dog" hunt that the Army organised near the front line in similar fashion to a foxhunt. Facing them were the Cossacks and when they saw all the activity in the British lines they became very agitated, assuming, reasonably enough, that the British were about to attack. After a while however, it dawned on their incredulous

The railway works at Balaclava, by night.

Progress of the Balaclava railway to the church of Kadikol (March 24, 1855).

Cossack minds, what was going on and they watched with bewildered fascination the progress of the hunt.

A little while after the British noticed two riders from the Cossack lines galloping towards them, so they detained them and found they were two deserters who had tired of fighting. The British officers found out that the two horses did not belong to the deserters so they took them back towards the Russian lines and pointing them towards the Cossacks, gave them a smack on their rump, and sent them back. Gentlemen did not steal horses.

Something similar occurred when a Russian deserter offered military information about the Russian defences. Lord Raglan flatly refused to listen to him because he was so appalled at the man's treason and sent him away with a flea in his ear.

Perhaps it was not the ideal way to win a war, but they have a certain appeal and incompetent though they were they were infinitely more likeable than some of the more efficient but cruel officers of other armies at that time.

We do not know what Brassey's navvies thought of the soldiers but a Captain Henry Clifford,

Balaclava 1855. Work on the railway is finished and supplies are unloaded from the ships in the harbour direct into the railway wagons and taken to the front at Sevastopol.

later Major General Sir Henry Clifford, wrote that the navvies looked "unutterable things" but did more work in a day than a regiment of soldiers did in a week. Not surprising since Beattie had not included "wild dog" hunting in the programme.

In just over six weeks the railway was finished and by the 7th April 1855 there were seventeen engines busy pulling the desperately needed supplies to the front, but from the very start of building the railway it was in use shortening the lines of supply and enabled the Army to survive that terrible winter. By the end of March they had carried 1,000 tons of shot and shell, 3,600 tons of clothing, blankets medical supplies and other goods.

In September 1855 Sevastopol fell and the great Russian naval base was destroyed by the victorious British Army still being supplied by the men of Peto, Brassey and Betts running the Crimean Railway.

Beattie, the engineer, had hardly any sleep for the three weeks before the convoy arrived, because he had to survey the route and do the prior planning and he worked continuously throughout the building of the railway. When it was finished, he was completely exhausted and worn out and although he was sent back to England he died shortly after his return. He gave his own life but he saved the lives of thousands of his fellow countrymen.

A correspondent wrote that the railway was expected to be finished by the end of April but was completed by mid March and commented, "the skill of the men entrusted with the building of the railway appears to have overcome all obstacles in a manner which few could have anticipated, even though accustomed to the celerity of workmanship in England and which to our soldiers in the Crimea, worn out by the failure of so many fine schemes, appeared an idle dream.

"It has once more proved that the men who have made England great by their skill and enterprise and powers of organisation, are of a far different calibre from the officials whom Government employs. While months have been spent in getting warm clothing and the barest necessities for hospital practice a few weeks only has been required for the conception and execution of a novel and most difficult enterprise."

Such was the power and skills of the Birkenhead firm of Peto, Brassey and Betts, Thomas Brassey and his men, they became national heroes and received acclaim in the Press and the Government heaped praise on them in the House of Commons, but unlike Florence Nightingale and others they are now almost forgotten—even in Birkenhead.

CANADA WORKS AND INDIA

For many years William Heap and Company did a considerable trade with India, supplying iron, steel, pumps and machinery and as in Canada, this business sprang from friendships forged while Mr Heap was working for Thomas Brassey on railway contracts in that country.

In 1855 the Governor General of India, Lord Dalhoussie made a minute in which this remarkable man planned the whole integrated network of railways to cover India and he even specified a uniform gauge of 5′–6″, the same as the Grand Trunk Railway, which was midway between the standard British Gauge of 4′–8½″ and the Great Western broad gauge of 7′–0″. The Indian Mutiny of 1857 accelerated the building of the railway as it encouraged the British Government to press ahead.

Victorian Imperialism is not fashionable in certain miserable quarters, but whilst agreeing that a people should rule themselves, Imperialism was inevitable with large masses of the earth backward and unexplored and given the need for Imperialism then history will show that it was fortunate that it was Britain that was cast in the role of chief Empire builder. By and large British rule was enlightened, and just as we benefitted from the Roman occupation, so India and other countries of the Empire benefitted from British rule. Even while the Indian Mutiny was in progress, the Universities of Calcutta, Madras and Bombay were founded, Grant in Aid helped the rapid expansion of private colleges, western style education was promoted on a large scale and the foundations of modern Indian medicine and science were laid.

Not all these measures were welcomed by the Indian ruling classes, the Hindus in particular did not like the suppression of suttee and infanticide. Even though the British stopped the widow being burned on her husband's funeral pyre, her family then treated her like a dog and she was made to wear sackcloth and ashes for the remainder of her life and was only allowed to do the most menial tasks in the household she had controlled up to the time of her husband's death.

Another great problem which had plagued India since time began was famine and this was one of the reasons why Lord Dalhousie wanted a railway network, because then foodstuffs could be rushed into famine areas. He also wanted to encourage new industries such as jute and cotton and good transport was needed to enable this to be done, but the Hindus were not enthusiastic because they felt that it would break down the Caste system.

It is true that the Mutiny also illustrated the need to move troops about quickly, but Lord Dalhousie had drawn up his plans two years earlier. It is fashionable in some quarters to draw a picture of a subjected people held down by brutal British force of arms, but this is quite untrue. There were occasions when there were only 20,000 British soldiers in the whole vast endless land mass of India with its teeming millions. Last year there were more troops in Northern Ireland than we had in the whole of India during our Imperial reign and we could not have ruled for so long without the consent of the people.

The Indian Civil Service was undoubtedly the most efficient, dedicated and uncorrupt in the East and this small band of 1,200 men gave wise Government to 300 millions. Today the Department of Industry alone employs in just a couple of its office blocks more people than were in the whole of the I.C.S. and they could not run a fish and chip shop. The I.C.S. laid down the legal and ruling system which the British bequeathed to India and in our day they had freedom of speech and if we had to imprison a comparative few, compare this with the situation now where the Socialist Mrs Gandhi is reported to have no less than 80,000 political opponents in jail including 30 M.P.s and freedom of speech is forbidden.*

It was not possible for Lord Dalhousie to obtain the capital to build the railways from the Indians and so all the cash was raised in Britain and the British Government guaranteed every mile of railway that was built. Eight firms were approached to build sections of the network and one of them was the greatest contractor of them all, Thomas Brassey and his men of Canada Works, Birkenhead.

EASTERN BENGAL RAILWAYS

In 1858 a year after the Mutiny, Thomas Brassey in partnership with Sir Joseph Paxton and Mr Wythes was awarded the contract for the 112 miles of the Eastern Bengal Railway. The engineer was Isambard Kingdom Brunel who not only built remarkable ships but engineered the splendid "Great Western Railway" known as Brunel's billiard table because the gradient was 1 in 1,380.

As usual, much was left to Thomas Brassey's men because while the Eastern Bengal was being built, Brunel was struggling with the immense problem of building the largest ship in the world the "Great Eastern". Indeed a year after work started Brunel died on Thursday, 15th September 1859, heartbroken when he learned that the feedwater heaters on the "Great Eastern" had blown up on the 9th September. He had pushed out, way beyond the then frontiers of engineering knowledge and it is sad that he died despairing with a feeling of failure.

The Eastern Bengal Railway ran from Calcutta to the village of Kushtia, further up the River Ganges and all the materials were made and despatched from England, much of it from Canada Works. On this contract lattice bridges were used rather than tubular, because it was easier to despatch material for a lattice design and William Heap was responsible for the bridges built on this railway.

The Eastern Bengal was not a financial success for Thomas Brassey but his appetite was whetted and he then tendered for the much larger Delhi Railway.

DELHI RAILWAY

The Delhi Railway was part of the East Indian Railway and runs 304 miles from Delhi to

*Written in 1976.

Umritsir in the Punjab. Brassey's contract included all the civil engineering work, the permanent way, station, locomotives, wagons, carriages and signalling equipment for the fixed price of £14,630 per mile. This was higher than the tender price for the Eastern Bengal, but Brassey had no intention of losing money twice.

Canada Works again made much of the equipment and the logistics were tremendous. Remember there were no roads, nor railways in this area at that time to carry the materials for the new railroad in India and the total weight of the material shipped out from England came to the staggering total of 100,000 tons and this had to be carried by a variety of transport, including ox carts, camels and elephants, over 1,000 miles from the landing point in India to the site of the railway.

BRIDGES OVER THE JUMNA RIVER

William Heap was given the job of building all the bridges on the Delhi line and there were many that were major engineering feats in themselves, including those over the River Sutley and the River Beeas, but the greatest of all the Indian bridges and one to rival the Victoria Bridge was the bridge over the Jumna River. Years later when William Heap was a Director of Canada Works Engineering Company and they invited subscriptions for the shares in the company, out of all the magnificent achievements of this superb works the prospectus gave as examples of the work that had been done there, the Victoria Bridge and the Jumna River Bridge.

The Indian rivers posed different problems to the St Lawrence because the width of the rivers varied tremendously from season to season and were quite unpredictable as to the course they would take. For example, the Jumna River at the end of the hot dry season is a placid stream but at the beginning of the spring when the Himalayan sun starts to melt the snow on the mountain ranges, small torrents start way up in the mountain, then join together to form tumbling rivers and finally these rivers join forces to become great belts of foaming, swirling, raging torrents where the railway had to cross the river.

The position chosen for the crossing was at Allahabad, about one and a quarter miles from its junction with the River Ganges and at that spot the depth of the river was only 15′, but three quarters of a mile away, up stream downstream, the depths increased to 65′ and 72′ respectively. Like all Indian rivers, the Jumna winds about a great deal in its course to the sea and the torrential flow of water when the snow starts to melt cut away the sand from the banks and scour out the bed of the river. The Ganges under the influence of the cutting power of the torrents has moved in places, one and three quarter miles east in the last forty-five years at the rate of 200 ft per year.

Great difficulties were experienced in preparing the foundations for the piers which were to support the fifteen spans and although Brassey's men went down to 80′ they only encountered sand with a few loose stones and so they made artificial islands of sand with sand bags and loose sand and on the islands they positioned ten iron curbs 13′–6″ outside diameter and 8′–6″ inside diameter and these were lowered into the sand by scooping out the sand from under, then another curb was set down on top and so on until they had a firm foundation. The masonry for the piers was then cemented to iron bars projecting up from the curbs.

The bridge was 3,278′ between abutments and consisted of fifteen openings of 205′ clear span, each crossed by wrought iron lattice girders thus giving a total waterway of 3,075′, the height to the top of the pier was 58′–6″ and the top of the rails is 80′ above low water level. At the last minute the River Jumna changed course again, so much so that a cable was sent to Canada Works saying that the length of the main viaduct would have to be increased to 6,600′ or one and a quarter miles long not counting the approach viaducts.

Eventually the bridges and the lines were finished and a report in "The Engineer" of 1862 on the opening of the railway stated "A Brahmin looking at the locomotive at Umritsir remarked 'All the incarnations of the Gods in India, never produced such a thing as that'. By this time the news had been carried by the trading caravans into Cabul and Central Asia and so our prestige increases."

THE CHORD LINE

The last railway line that William Heap worked on in India was also Brassey's last Indian contract; this was the Chord Line which ran 147 miles on the Delhi to Calcutta, East Indian Railway. It was called the Chord because it cut off a great semi-circular loop on this line.

THE HERITAGE OF THE INDIAN RAILWAYS

The nervous system of modern India is the railway network, it has done everything that Lord Dalhousie, that man of vision, planned. It reduced the pestilance of famine, it enabled the Indian cotton, jute and tea industries to be set up and prosper and, as the Hindus feared, it did help to break down the caste system. Of this the people of Birkenhead have much to be proud, because of the important part played by Canada Works.

As well as having a long term economic benefit the building of the railway system in India brought some immediate benefit to the Indian workers employed in building the railway. Much of the British capital spent on the railway was paid to the Indian workers, labouring under the supervision of the British navvies. This was referred to by Mr Henfrey, Thomas Brassey's partner in the Delhi Railway at the opening of the Meerut and Umballa section of the Delhi Line, when he said "How greatly the working classes of this country (India) have profitted by the construction of the railways may be judged by the fact that out of seventy-five to eighty million pounds sterling expended to this present time on Indian railways nearly two thirds, or between forty to fifty million pounds has passed, I cannot say into the pockets, but into the hands of the working classes."

When George V came to the throne in 1910 as Emperor of India he was able to see a railway network in India comprising 31,500 miles of track and every mile of track had been financed by British capital and built by British firms. This may be compared with the railway network in present day Germany which totals 18,000 miles and which is about to be reduced to 10,000 miles or with the total motorway mileage built in the U.K. which is 1,312 miles to date. In the whole of the vast sub-continent of India there was fast transport and in the vast, densely populated, band of country from Calcutta to Peshawe it was hard to find a town that was more than 25 miles from the railway.

When our splendid Imperial past is under attack, compare the British achievements in India with the state of a similar country facing the same sort of problems, enormous population, no birth control, poor economy, large land mass—China.

In that country whole densely populated provinces 400 miles across had no railway system, nor metalled roads and in that gigantic, endless, country there was only 3,000 miles of railway.

CALCUTTA WATERWORKS

Another of Lord Dalhousie's schemes was to improve irrigation and water supply throughout India and the British built irrigation canals and piped water into the teeming cities just as two thousand years earlier the Romans laid wooden pipes from the wells at Boughton into the centre of Chester. When in 1820 water flowed into the centre of Delhi the population went wild with excitement, donned gala dress and there were great celebrations in the streets.

Another city with enormous water problems was Calcutta and in 1865 it was decided to build a completely new water works system. The overall plan was done by the Chief Engineer for Calcutta, William Clark. He designed the three large pumping stations, the filter beds, settling tanks and distribution network.

Thomas Brassey was awarded this contract, but ill health forced William Clark to return temporarily to England before the scheme started. William Heap was given this contract to handle because the Bridge Department at Canada Works were specialists in pumping because of the numerous coffer dams they had to pump out. William Heap became very friendly with William Clark and in 1872, Mr Clark seconded Mr Heap's application for Membership of the Institution of Mechanical Engineers.

William Clark was formerly a resident engineer with the East Indian Railway Company and when later he became Engineer to the Municipality of Calcutta he carried out another very important scheme by designing a completely new sewerage scheme for the city, but there was very strong local objections and he had to fight hard to push the scheme through.

William Clark remained as Chief Engineer for the City of Calcutta until 1874 when he finally retired to England and entered into Partnership with W. F. Batho who proposed William Heap's application for the Institute of Mechanical Engineers. Together they designed a number of machines including the invention of the first steam driven road roller.

The friendships William Heap formed in India during the building of the railways and Calcutta Waterworks stood him in good stead when he set up on his own and for many years we supplied pumps, machinery and wrought iron pontoons to Calcutta Municipal Authority, steel for bridges, railway equipment and high grade iron and steel to Jessops of Calcutta who were the Purchasing Agents for the Indian Government.

AUSTRIAN RAILWAY AND CANADA WORKS

At the same time as the Indian Railways were being built Thomas Brassey was building in many other countries. In 1863 Thomas Brassey was awarded the 165 mile Lemberg and Czernowitz Railway in Austria and its story illustrates the sort of man that worked for Thomas Brassey.

Apart from the technical and logistical problems to be overcome, Mr Brassey had great financial problems; the skies were darkening and the sound of distant thunder heralded the build up to the financial storms of 1866. Another small difficulty his men had to contend with was that Austria and Prussia were at war and their armies were fighting fiercely on the sector of the Crocow and Lemberg railway line which Peto, Brassey and Betts had recently finished.

At one point money was running out for the firm in Austria and it was necessary to get cash up the line which was closed because of the battle. Mr Ofenheim, Thomas Brassey's agent decided that if they waited until the battle was finished it might be too late for the firm and so he decided that he must do something about it. He found one of the Birkenhead-built locomotives in a shed and then fired it up and built up a head of steam and set off down the line. A medieval fight was about to be interrupted by the Industrial Revolution.

As Mr Ofenheim looked out of the locomotive he could see that he was approaching the battlefield, there were wounded stragglers dragging themselves along and wild, fear crazed, horses galloping away. A fierce fight was taking place with cavalry engaged in bloody hand to hand fighting, cannons were booming and men and horse screamed as fragments of iron tore into their bodies. Mr Ofenheim gave a mighty blast on the locomotives whistle which startled men and horses on both sides. The cavalry struggled to control their horses, the gunners with lighted tapers in their hands stood open mouthed at the astonishing sight of such a monster as they had never seen before charging towards them, smoke belching from its chimney, at the unprecedented speed of 50 m.p.h. They all stood motionless for a few seconds and then realising that the monster was not going to stop they scattered away from the railway line and stood watching the progress of the locomotive with awe and perhaps a little admiration. It took more than a battle to stop Thomas Brassey's men going about their business.

This feat was relayed by the Austrian officers present to their Emperor, Franz Joseph of Austria who was so impressed that he sent for Victor Ofenheim, and after congratulating him asked: 'Who is this Mr Brassey, this English contractor, for whom men are to be found who will work with such zeal and risk their lives?"

For his work in Austria the Emperor awarded Thomas Brassey the Cross of the Iron Crown, but one suspects that Mr Brassey valued Mr Ofenheim's loyalty and bravery more than medals.

THE CRASH OF 1866

The year 1866 saw the culmination of a crisis for Thomas Brassey and the other railway contractors which was to lead directly to the formation of our company. The crisis had been gathering momentum for some time and was eventually triggered off by the failure of the London private bank of Overend & Gurney, for forty years one of the World's greatest financial houses.

A year earlier in 1865, the Indian cotton market built up under Lord Dalhousie's guidance, collapsed and a number of small failures occurred which were symptomatic of the underlying financial weakness, so that when Overend & Gurney went under with financial liabilities of nineteen million pounds, it caused a shock wave of bankruptcies all over the country. One of the first to fail was Sir Morton Peto, Thomas Brassey's old partner became almost all his wealth had been tied up with Overend & Gurney. There was a run on the banks and on Friday, 11th May 1866, Mr Gladstone, the Chancellor of the Exchequer raised Bank Rate to 10% and credit became unobtainable. The shock waves then spread out from this country and companies all over the world collapsed and only the strongest survived.

Thomas Brassey rode out the storm because he finished the Lemberg and Czernowitz Railway just in time and this released one million pounds, he used part of his own private fortune and although credit was almost unobtainable his standing was so high that he managed to borrow another quarter of a million pounds.

It became apparent to William Heap that although Peto, Brassey and Betts had survived the crisis, the great heyday of railway building was passing, he had been with Thomas Brassey for almost fourteen years, ten of those as the Manager of the Bridge Department of Canada Works and during that time had taken part in some of the greatest engineering adventures ever undertaken. He had met, and worked with, Robert Stephenson, Isambard Kingdom Brunel, Thomas Brassey, Sir William Fairbairn, Eaton Hodgkinson, Joseph Locke and others, but now he felt that at the age of forty-years, he should start up his own business.

William Heap's parting with Thomas Brassey and George Harrison was very friendly and indeed until the firm finally went under after the death of Mr Brassey we were favoured with a considerable volume of business with Peto, Brassey and Betts and later when Canada Works became a public company, William Heap was invited to be a Director.

THE END OF THOMAS BRASSEY

William Heap's judgement proved sound because after he had left his old firm they only built another 550 miles of railway and in the spring of 1870 Thomas Brassey was told that he had cancer. He refused to stop working and continued to visit sites in Britain checking that the work was being done well and looking to the welfare of his men.

In the late summer of 1870, Mr Brassey was forced to take to his bed and in those days suffering was more intense than today because there were no pain killing drugs. Right up to the end his engineers and navvies used to visit him, some of them walking great distances, so they could quietly pay their last respects.

Finally on 8th December 1870, Thomas Brassey, the first great contractor in the world died, a gentle, soft spoken, kind man, a brilliant engineer and an organising genius. He had built 1 out of every 20 miles of the world's railways, even as early as 1850 he had built 3 out of every 4 miles of French railways, he had helped to open up continents and changed social pattern all over the world.

Thomas Brassey left £3,200,000 and he deserved every penny of it, he left £2,000,000 on trust for his children, but more important he left a heritage of railways all over the world and to this day he is remembered with love, affection and admiration by the few who know of him.

In the Birkenhead News of 21st August 1886 there was a letter from John Wilson of Congleton who wrote about the granting of a Peerage to Thomas Brassey's son Thomas, Lord Brassey "Whilst Thomas Brassey cared for no distinction for himself he would have been proud of his son's well deserved Peerage, but he would have been most proud of his son's words the other day; 'In being allowed to choose my future title I wish for no name but that of my father to whom I owe everything'".

In his lifetime Thomas Brassey gently, but firmly, refused all honours offered in this country, but he accepted those offered by foreign countries because he thought that a refusal might cause offence. Having lost the Legion of Honour bestowed on him by the Emperor Napoleon III of France and the Order of St Maurice and St Lazarus bestowed on him by King Victor Emanuel of Italy, he sent out for duplicates to please his wife and as he remarked "Mrs Brassey will be glad to possess all these crosses".

Today the Brassey family live in Apthorne Hall near Oundle in Northamptonshire; the first Lord and Lady Brassey were remarkable in their own way because they sailed around the world in their, Mersey built, yacht "Sunbeam" and Lord Brassey wrote a book about the classical doctrine of supply and demand, and the present members of the family are also not without distinction.

In 1966 the late Lord Brassey bought a small circular hover machine to raise money for the local church and he became so interested in hovercraft that he founded a firm "Hoverair" to make this type of vehicle. Lord Brassey died in the Autumn of 1966 but the amazing Lady Brassey carried on with the work and by the end of 1968 had produced a magnificent little Hovercraft, one of which was used in 1969 by the young adventurer Captain Sir Ranulph Twisleton-Wykeham-Fiennes in his expedition to the source of the White Nile.

BIOGRAPHY OF THOMAS BRASSEY

In 1872 Sir Arthur Helps, Clerk of the Privy Council, wrote "The Life and Labours of Mr Brassey (1805–1870) and he wrote the following dedication to Queen Victoria:

"To the Queen.

Madame,

I am very grateful for the permission given me to dedicate this work to your Majesty.

I desire so to dedicate it, because I do not know of any one who has a deeper sympathy with the labouring classes than Your Majesty, or anyone who takes a more heartfelt interest in everything that concerns their habits, their education and their general welfare. Moreover this sympathy, and this interest are not confined to those classes in Your Majesty's Dominions only, but are extended to them wherever they are to be found.

I think also that it cannot but be very gratifying to Your Majesty to have full evidence that, in a special kind of labour of a very important character, namely the construction of railways, your own subjects have hitherto borne the palm, and have introduced their excellent modes of working into various Foreign countries.

Your Majesty will find that the late Mr Brassey was an employer of labour after Your Majesty's own heart, always solicitous for the wellbeing of those who served under him, never keeping aloof from them, but using the powerful position of Master in such a manner as to win their affection and to diminish the distance which is often too great between the employer and the employed.

I venture, therefore, to think that the volume will be of interest to Your Majesty on its own account; and that you will be disposed to view with favour the merits, if any, and to deal gently with the faults of a work written by one who, with all respect, is ever Your Majesty's.

Faithful and devoted
Subject and servant
Arthur Helps.

London, June 1872.

Sir Arthur Helps gave a list of Thomas Brasseys contracts to which we have added one or two others as follows:

TABLE
OF
RAILWAY AND OTHER CONTRACTS

COMPLETED BY MR BRASSEY BETWEEN THE YEARS 1834 AND 1870

Year	Contract	Partner	Engineer	Agent	Mileage
1834	Branborough Road				4
1835	Grand Junction Railway		Mr Locke, M.P., F.R.S.		10
1837	London and Southampton Railway (Branches and Maintenance)		Mr. Locke, M.P., F.R.S., Mr Neuman	Mr Ogilvie and others	36
1839	Chester and Crewe Railway		Mr R. Stephenson, M.P., F.R.S.	Mr G. Meakin	11
	Glasgow, Paisley, and Greenock Railway		Mr Locke, M.P., F.R.S.	Mr Strapp and others	7
	Sheffield and Manchester Railway		Mr Locke, M.P., F.R.S.	Mr Dent and others	19
1841	Paris and Rouen Railway	Mr Mackenzie	Mr Locke, M.P., F.R.S., Mr Neuman	Mr E. Mackenzie, Mr J. Jones, Mr Goodfellow, Mr Day and others	82
1842	Orleans and Bordeaux Railway	Messrs. W. & E. Mackenzie	M. Pepin Lehalleur	Mr E. Mackenzie	294
1843	Rouen and Havre Railway	Mr Mackenzie	Mr Locke, M.P., F.R.S., Mr Neuman	Mr Day, Mr C. Smith, Mr J. Jones, Mr Swanson, Mr Goodfellow	58
1844	Lancashire and Carlisle Railway	Mr Mackenzie, Mr Stephenson	Mr Locke, M.P., F.R.S.	Mr G. Mould	70
	Colchester and Ipswich Railway	Mr Ogilvie	Mr Locke, M.P., F.R.S., Mr Bruff	Mr Ogilvie	16
	Amiens and Boulogne Railway	Mr W. Mackenzie, Mr E. Mackenzie	M. Bazaine, Sir W. Cubitt, F.R.S.	Mr E. Mackenzie	53
1843	Trent Valley Railway	Mr Makenzie, Mr Stephenson	Mr R. Stephenson, M.P., F.R.S., Mr Bidder, Mr Gooch	Mr J. Jones, Mr S. Horn	50
	Chester and Holyhead Railway	Mr Mackenzie, Mr Stephenson	Mr R. Stephenson, M.P., F.R.S., Mr Ross, Mr F. Forster	Mr Woodhouse	31
	Ipswich and Bury Railway	Mr Ogilvie	Mr Locke, M.P., F.R.S., Mr Brutt	Mr Ogilvie	24½
	Kendal and Windermere Railway	Mr Mackenzie, Mr Stephenson	Mr Locke, M.P., F.R.S.	Mr G. Mould	12
	North Wales Mineral Extension Railway	Mr Mackenzie, Mr Stephenson	Mr Robertson	Mr Meakin	5
1845	Caledonian Railway (1st Contract)	Mr Mackenzie, Mr Stephenson	Mr Locke, M.P., F.R.S.	Mr G. Mould, Mr Woodhouse	125
	Clydesdale Junction Railway	Mr Mackenzie, Mr Stephenson	Mr Locke, M.P., F.R.S.	Mr Woodhouse	15
	Greenock Harbour	Mr Mackenzie, Mr Stephenson	Mr Locke, M.P., F.R.S.	Mr Goodfellow, Mr Barnard	—
	Scottish Midland Junction Railway	Mr Mackenzie, Mr Stephenson	Mr Locke, M.P., F.R.S.	Mr G. Mould	33
	Scottish Central Railway	Mr Mackenzie, Mr Stephenson	Mr Locke, M.P., F.R.S.	Mr Falshaw	46
1846	Lancashire and Yorkshire Railway (Maintenance)	Mr Field	Mr Hawkshaw, F.R.S.	Mr Day	93
	Ormskirk Railway	Mr Mackenzie	Mr Meek	Mr Greene	30
	Shrewsbury and Chester Railway	Mr Mackenzie, Mr Stephenson	Mr Robertson	Mr Meakin	25
	Mineral Line (Wales)	Mr Mackenzie, Mr Stephenson	Mr Robertson	Mr Meakin	6½
1847	Buckinghamshire Railway		Mr Dockray	Mr S. Horn	47½
	Birkenhead and Chester Junction Railway		Mr Rendel, F.R.S.	Mr Goodfellow, Mr Day	17½

Year	Contract	Partner	Engineer	Agent	Mileage
	Haughley and Norwich Railway	Mr Ogilvie	Mr Locke, M.P., F.R.S.	Mr P. Ogilvie	33
	Great Northern Railway		Sir W. Cubitt, F.R.S.	Mr Bartlett, Mr Milroy, Mr Ballard	75½
	North Staffordshire Railway		Mr R. Stephenson, M.P., F.R.S., Mr Bidder	Mr J. Jones	48
	Shrewsbury Extension Railway	Mr Mackenzie, Mr Stephenson	Mr Robertson	Mr Meakin	3
	Trent Valley Stations	Mr Mackenzie, Mr Stephenson	Mr Bidder and others	Mr Holme	—
	Blackwall Extension Railway	Mr Ogilvie	Mr Locke, M.P., F.R.S., Mr Stanton	Mr Burt	1¾
	Richmond and Windsor Railway	Mr Ogilvie	Mr Locke, M.P., F.R.S.	Mr Evans	16½
	Rouen and Dieppe Railway	Mr Mackenzie	Mr Neuman, Mr Murton	Mr Benyon, Mr C. Smith	31
1848	Chester Station			Mr S. Holme	—
	Oswestry Branch Railway	Mr Mackenzie, Mr Stephenson	Mr Robertson	Mr Meakin	2
	Loop Line (L. & S.W.R.)	Mr Ogilvie	Mr Locke, M.P., F.R.S.	Mr Evans	7
	Caledonian Railway (2nd Contract) (Stations, Maintenance, etc.)	Mr Mackenzie,	Mr Locke, M.P., F.R.S.		—
	Glasgow and Barhead Railway		Mr Locke, M.P., F.R.S.	Mr Strapp	11
	Barcelona and Mataro Railway	Mr Mackenzie	Mr Locke, M., F.R.S. Mr W. Locke	Mr Robson	18
1849	Royston and Hitchin Railway		Mr Locke, M.P., F.R.S.	Mr H. Harrison	13
1850	Shepwreth Extension Railway		Mr Locke, M.P., F.R.S.	Mr H. Harrison	5
	Birkenhead Docks		Mr Rendel, F.R.S., Mr Abernethy	Mr Dent	—
	North and South-Western Junction Railway	Mr Ogilvie	Mr G. Berkeley	Mr Evans	4
	Prato and Pistoja Railway		Italian Government	Mr T. Woodhouse	10
1851	Shrewsbury and Hereford Railway, and Maintenance		Mr Robertson	Mr W. Field	51
	Norwegian Railway	Sir M. Peto, M.P., Mr Betts	Mr Bidder	Mr Merrit, Mr Earle	56
1852	Hereford, Ross, and Gloucester Railway	Sir M. Peto, M.P., Mr Betts	Mr Brunel, F.R.S.	Mr Watson	30
	London, Tilbury, and Southend Railway	Sir M. Peto, M.P., Mr Betts	Mr Bidder	Mr White	50
	Victoria Docks and Warehouses	Sir M. Peto, M.P., Mr Betts	Mr Bidder	Mr Holland	—
	Warrington and Stockport Railway		Mr Lister	Mr Goodfellow	12
	North Devon Railway	Mr Ogilvie	Mr W. R. Neale	Mr P. Ogilvie, Mr Evans	47
	Mantes and Caen Railway		Mr Locke, M.P., F.R.S., Mr Neuman, Mr W. Locke	Mr J. Jones, Mr C. Jones, Mr J. Milroy	113
	Le Mans and Mezidon Railway		Mr Locke, M.P., F.R.S., M. Bergeron	Mr Woodhouse	84
	Lyons and Avignon Railway	Sir M. Peto, M.P., Mr Betts	M. Talabot, M. Thirion, M. Molard	Mr G. Giles, Mr Murton	67
	Dutch Rhenish Railway		Mr Locke, M.P., F.R.S.	Mr Ballard	43
	Grand Trunk Railway	Sir, M. Peto, M.P., Mr Betts, Sir W. Jackson	Mr Ross	Mr Reckie, Mr Hodges, Mr Rowan, Mr Tait, Mr P. Ogilvie	539
1853	Crystal Palace and West-End Railway	Sir M. Peto, M.P., Mr Betts	Mr Bidder	Mr Watson	5
	Sambre and Meuse Railway		M. Declerq	Mr H. Harrison	28
	Turin and Novara Railway		Italian Government	Mr T. Woodhouse, Mr Hancox	60
	Hauenstein Tunnel		M. Etzel	M. Benyon	1½
	Royal Danish Railway	Sir M. Peto, M.P., Mr Betts	Mr Bidder, Mr G. R. Stephenson	Mr McKeon	75
1854	Arpley Branch Railway		Mr Lister	Mr Goodfellow	1½
	Woodford and Loughton Railway		Mr Bidder	Mr H. Harrison	7½
	Central Italian Railway	Sir W. Jackson, Mssrs. Fell and Jopling	Italian Government	Mr Fell, Mr Jopling	52
	Turin and Susa Railway	Mr C. Henfrey	Italian Government	Mr C. Henfrey	34
	Bellegarde Tunnel	Messrs. Parent and Buddicom	M Talabot	Mr Goodfellow	2½
1855	East Suffolk Railway	Sir M. Peto, M.P., Mr Betts	Mr G. Berkley	Mr Watson	63

Year	Contract	Partner	Engineer	Agent	Mileage
	Inverness and Nairn Railway	Mr Falshaw	Mr Mitchell	Mr Falshaw	16
	Portsmouth Direct Railway	Mr Ogilvie	Mr Locke, M.P., F.R.S.	Mr Evans	33
	Caen and Cherbourg Railway		Mr Locke, M.P., F.R.S., Mr W. Locke	Mr Milroy, Mr C. Jones	94
	Coghines Bridge		Italian Government	Mr Dent	—
1856	Woodbridge Extension Railway	Mr Ogilvie	Mr Bruff	Mr Boys	10
	Elizabeth-Linz Railway	Sir M. Peto, M.P., Mr Betts	M. C. Keissler	Mr G. Giles	49
1857	Leicester and Hitchin Railway		Messrs. Liddell and Gordon	Mr Horn, Mr Harrison	62½
	Leominster and Kington Railway	Mr Field	Mr Wylie	Mr Field	14
	Minories Warehouses	Sir M. Peto, M.P., Mr Betts	Mr Tite, M.P.	Mr Holland	—
1858	Leatherhead, Epsom, and Wimbledon Railway	Mr Ogilvie	Mr Locke, M.P., F.R.S., Mr A. C. Crosse	Mr Ogilvie	10
	Worcester and Hereford Railway	Mr Ballard	Messrs, Liddell and Gordon	Mr Ballard	26
	Inverness and Aberdeen Junction Railway	Mr Falshaw	Mr Mitchell	Mr Falshaw	22
	Bilbao and Miranda Railway	Messrs. Wythes, Paxton and Bartlett	Mr Vignoles, F.R.S.	Mr Bartlett	66
	Eastern Bengal Railway	Messrs. Wythes, Paxton and Henfrey	Mr Hawkshaw, F.R.S.	Mr C. Henfrey	112
1859	Cannock Mineral Railway	Mr Field	Mr Addison	Mr J. Stephenson	10
	Crewe and Shrewsbury Railway		Mr Locke, M.P., F.R.S.	Mr Day	32
	Salisbury Station	Mr Ogilvie	Mr Errington, Mr Tolme	Mr Caswell	2
	Denny Branches		Mr Locke, M.P., F.R.S., Mr Errin-gton	Mr Falshaw	3
	Victor Emmanuel Railway	Sir W. Jackson, Mr Henfrey	Mr Neuman, Mr Ranco	Mr Bartlett, Mr W. St-rapp, Mr Blake, Mr Ed-wards	73
	Ivrea Railway	Mr C. Henfrey	Italian Government	Mr Dent, Mr Dixon	19
	Great Northern, Great Eastern, Great Southern Railways (New South Wales)	Sir M. Peto, M.P., Mr Betts	Mr Whitton	Mr Wilcox, Mr Rhodes	54
1860	Salsbury and Yeovil Railway	Mr H. Harrison, Mr Ogilvie	Mr Locke, M.P., F.R.S.	Mr H. Harrison	40
	Woofferton and Tenbury Railway	Mr Field	Mr Wylie	Mr Mackay	5
	Wenlock Railway	Mr Field	Mr Fowler	Mr Seacome	4
	Port Partick Railway	Mr Falshaw	Mr Blyth	Mr Falshaw	17
	Stokes Bay Pier and Branch Railway	Mr Ogilvie	Mr Fulton	Mr Evans	2
	Harleston and Beccles	Mr Ogilvie	Mr Bruff	Mr Boys	13
	Dieppe Railway (laying second road)	Mr Buddicom	M. Julien	Mr R. Goodfellow	—
	The Maremma, Leghorn, etc., Rail-way		M. Pini	Mr Jopling, Mr C. Jones	138
	Jutland Railway	Sir M. Peto, M.P., Mr Betts	Danish Government	Mr Rowan	270
1861	Disley and Hayfield Railway	Mr H. Harrison	Mr Errington	Mr Harrison	3½
	Knighton Railway	Mr Field	Mr Robertson	Mr Field	12
	Nuneaton and Hinckley Railway	Mr Field	Mr Addison	Mr J. Stephenson	5
	Shrewsbury and Hereford Railway (widening)	Mr Field	Mr Wylie, Mr Clark	Mr J. Mackay	51
	West London Railway (extension)	Mr Ogilvie	Mr W. Baker	Mr Evans	9
	Ludlow Drainage	Mr Field	Mr Curley	Mr Mackay	7
	Severn Valley Railway		Mr Fowler	Mr Field, Mr Day, Mr Dent, Mr Dowell	42
	South Staffordshire Railway	Mr Field	Mr McClean, M.P., F.R.S.	Mr Day	4
	Metropolitan Mid Level Sewer	Mr Ogilvie, Mr Harrison	Mr Bazalgette, C.B.	Mr Harrison	12
1862	Ringwood and Christchurch Railway	Mr Ogilvie	Captain Moorsom	Mr Evans	8
	Kingston Extension Railway	Mr Ogilvie	Mr Galbraith	Mr Evans	4
	Cannock Chase Railway	Mr Field	Mr Addison	Mr Cooper	3
	Coalbrookdale Railway	Mr Field	Mr Fowler	Mr Dent	5
	Ashchurch and Evesham Railway	Mr Ballard	Messrs. Liddell and Gordon	Mr Ballard	11
	Nantwich and Market Drayton Rail-way	Mr Field	Mr Gardiner	Mr Gallaher, Mr Mac-kay	11
	South Leicester Railway	Mr Field	Mr Addison	Mr J. Stephenson	10
	Tenbury and Bewdley Railway	Mr Field	Mr Wylie, Mr Clarke	Mr Mackay	12

Year	Contract	Partner	Engineer	Agent	Mileage
	Wenlock and Craven Arms Railway	Mr Field	Mr Fowler	Mr Dent, Mr N. Mackay	14
	Ludlow and Clee Hill Railway	Mr Field	Mr Wylie, Mr Clarke,	Mr Mackay	6
	Llangollen Railway		Mr Robertson, M.P.	Mr Gallagher	6
	Rio Janeiro Drainage	Mr Ogilvie	M. Gotto	Mr Honcox	—
	Mauritius Railway	Mr Wythes and others	Mr Hawkshaw, F.R.S.	Mr Longridge	64
1863	Epping and Ongar	Mr Ogilvie, Mr Harrison	Mr Sinclair	Mr Harrison	13
	Barrow Docks	Mr Field	Mr McLean, M.P., F.R.S.	Mr Dent	—
	Runcorn Branch Railway	Mr Ogilvie	Mr W. Baker	Mr Evans	9
	Tendring Hundred Railway	Mr Ogilvie	Mr Bruff	Mr Boys	3
	Worm Drainage	Mr Field	Mr Curley	Mr Mackay	—
	Sudbury, Bury St Edmunds, and Cambridge Railway	Mr Ogilvie	Mr Sinclair	Mr Bell, Mr Boys, Mr Smalls	48
	Meridionale Railway	M. Parent, Mr Budicom	M. Grattoni and others	Mr C. Jones, Mr Charles	160
	Queensland Railway	Sir M. Peto, M.P., Mr Betts	Mr Fitzgibbon	Mr Wilcox	$78\frac{1}{4}$
	North Schleswig Railway	Sir M. Peto, M.P., Mr Betts	Mr Rowan	Mr Louth	70
1864	Epping Railway	Mr H. Harrison, Mr Ogilvie	Mr Sinclair	Mr Butler	12
	Letton Drainage	Mr Field	Mr Curley	Mr Mackay	3
	Dunmow Railway	Mr Ogilvie, Mr H. Harrison	Mr Sinclair	Mr Butler	18
	Corwen and Bala Railway		Mr Robertson, M.P.	Mr Field, Mr Reid	14
	Wellington and Market Drayton Railway	Mr Field	Mr Wilson	Mr Mackay	16
	Enniskillen and Bundoran Railway	Mr Field	Mr Hemens	Mr Day, Mr Drennan	36
	Central Argentine Railway	Mr Wythes, Mr Wheelwright, Mr Ogilvie	Mr E. Woods	Mr Wheelwright	247
	Lemberg Czernowitz Railway		Mr McClean, Mr Stileman, M. Ziffer, M. de Herz	Mr Strapp	165
	Viersen-Venlo Railway	Mr Murton	M. Lange, Mr Bidder	Mr Murton	11
	Delhi Railway	Mr Wythes, Mr Henfrey	Mr J. Harrison, Mr Bidder	Mr C. Henfrey, Mr Mareiller	304
1865	Chertsey Extension Railway	Mr Harrison	Mr Galbraith	Mr Harrison	$3\frac{1}{2}$
	Dee Reclamation Works	Mr Field Mr Meakin	Mr Bateman, F.R.S.	Mr Meaken	—
	Evesham and Redditch Railway	Mr Ballard	Mr Richards	Mr Ballard	18
	East London Railway	Mr Wythes, Messrs. Lucas, Brothers	Mr Hawkshaw, F.R.S.	Mr H. Harrison	$2\frac{1}{2}$
	Hull and Doncaster Railway	Mr Field	Mr Harrison	Mr Stephenson	16
	Hereford Loop Railway	Mr Field	Mr Clarke	Mr Mackay	$2\frac{1}{2}$
	Hooton and Parkgate Railway	Mr Field	Mr Johnson	Mr Mackay	5
	London and Bedford Railway	Mr Ballard	Mr Liddell	Mr Ballard	36
	Llangollen and Corwen Railway	Mr Field			10
	Nantwich and Market Drayton (Widening)	Mr Field	Mr Gardiner	Mr Reid, Mr Mackay	11
	Boca and Barracas Railway	Mr Wythes, Mr Wheelwright	Mr Coghlan	Mr Simpson	3
	Warsaw and Terespol Railway	Mr Vignoles, F.R.S.	Russian Government	Mr H. Vignoles	128
	Chord Line (India)	Mr Wythes, Mr Perry,	Mr Rendel	Mr Perry	147
	Calcutta Waterworks	Mr Wythes, Mr Aird	Mr Purdon, Mr Lewis	Mr Paton	—
	Runcorn Bridge (Masonry only)				
1866	Ebbw Vale Railway	Mr Field	Mr Gardiner	Mr Mackay	2
	Thames Embankment	Mr Ogilvie, Mr Harrison	Mr Bazalgette, C.B.	Mr Harrison	$\frac{3}{4}$
	Kensington and Richmond Railway (and Spurs)	Mr Ogilvie. Mr Harrison	Mr Galbraith, Mr Tolmé	Mr Evans	7
	Christchurch and Bournemouth Railway	Mr Ogilvie	Mr Strapp	Mr Ogilvie	4
	Moreton Hampstead Railway	Mr Ogilvie	Mr Margrafy	Mr Crossley	12
	Bala and Dolgelly Railway		Mr Wilson	Mr Field, Mr Day, Mr Drennan	18
1867	Sirhowy Railway	Mr Field	Mr Sayer	Mr Mackay	2
	Wolverhampton and Walsall Railway	Mr Harrison, Mr Ogilvie	Mr Addison	Mr Harrison	$7\frac{1}{2}$
	Czernowitz Suczawa Railway		M. Ziffer, M. de Herz	Mr Strapp	60
	Kronprinz Rudolfsbahn	M. Klein, M. Schwarz	M. F. Kagda	M. Fölsch	272

Year	Contract	Partner	Engineer	Agent	Mile-age
1868	Silverdale Railway		Mr Forsyth	Mr Field, Mr Mackay	$13\frac{1}{2}$
1869	Nepean Bridge	Sir M. Peto, M.P., Mr Betts	Mr Fowler	Mr Willcox	—
1870	Callao Docks		M. Alléon	Mr Hodges	—
	Vorarlbergbahn	M. Klein, M. Schwarz	M. W. Paravicini	M. Fölsch	55
	Suczawa and Jassy Railway		M. Ziffer	Mr Strapp, Mr Edwards	135

CANADA WORKS AFTER THOMAS BRASSEY

Canada Works continued in business after Thomas Brassey's death, building locomotives, bridges, boilers and ships and in 1877 they built some of the floating pontoons of the Princes Landing Stage, Pierhead, Liverpool and one of the hinged walkways linking the landingstage to the river wall. The workmanship and ornateness of this walkway is superior to others of the same period.

In February 1886 "The Canada Works Engineering and Shipbuilding Company" was floated as a limited liability company and William Heap was one of the Directors of the new company. The prospectus quotes the bridges over the Jumna River on the East India Railway and the Victoria Bridge in Canada as examples of the class of work Canada Works had accomplished in the past.

Bust of Thomas Brassey in Chester Cathedral.

The new venture was not a success and the new company finally closed down in 1889. William Heap and Co. were owed £900 but Canada Works paid out 15/- in the pound and later paid off most of the remainder.

A sad end to a great engineering works, but it did not die completely because when it went under we took over many of the old connections that had been forged by Thomas Brassey and to this day we still benefit.

Today there is little to be seen of Canada Works except the entrance gates and one or two of the smaller buildings that still stand. In 1972 part of the boulder road that ran through the works was relaid in the gardens in front of our Hoylake office and a few cobbles were set in the paving in the garden around our Eccles office as a memorial to a great engineering works. Today the site is derelict and overgrown; a dumping ground for rubbish.

When one stands on the site of Canada Works one can hear the first triumphant whistle of the "Lady Elgin" in 1854; the singing of the workmen and the chinking of their glasses as they enjoyed their party at the bottom of the coffer dam in the River St Lawrence; the cheers of the workmen as the "Wildfire" sailed for Balaclava; the creaking of the carts and the cries of the drovers as they carried the 100,000 tons of equipment through the dust and the heat of the Indian plains; the shriek of the locomotive as Victor Offenheim charged through the warring Prussian and Austrian armies, and a quiet voice saying "Mrs Brassey will be glad to possess all these crosses".

NOTE TO THE THIRD EDITION 1991

A strange coincidence occurred one Saturday afternoon in 1980 when the writer received a telephone call from an unknown reader in Derby, who had just returned from his daughter's school fete. On a jumble stall he had seen a box of Victorian drawing instruments that had belonged to a John Millar (no relation to the writer so far as is known), who had been a draughtsman at H.M. Dockyard, Pembroke Dock, for many years until he retired about 1910 and he asked if we were interested. The writer bought the drawing instruments and found under the top layer of instruments a visiting card belonging to Sir Thomas Brassey, Thomas Brassey's son, who later became Lord Brassey.

BIRKENHEAD'S MEMORIAL

In his lifetime Thomas Brassey was lionised as a hero, he was fêted and decorated by kings and emperors and if he had wished he could have had honours heaped on him by his own country. His children restored St Erasmus' Chapel in Chester Cathedral and there is a bust of Thomas Brassey on the North Wall, but where is the reminder in Birkenhead of this great man and his superb works? A street is named after him and there is a seedy pub called "The Grand Trunk" but that is all.

The men who worked for Peto, Brassey and Betts were Victorian giants and they deserve to be remembered as does their leader. Birkenhead's past is not so illustrious that it can afford to ignore them and their deeds.

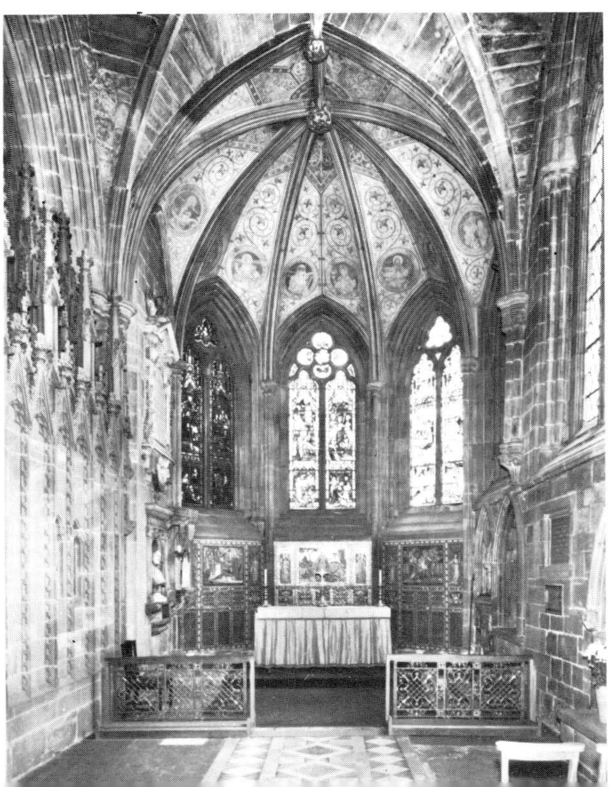

St. Erasmus Chapel, Chester Cathedral with Brassey's bust on the left. Thomas Brassey's children paid for the restoration of the Chapel as a memorial to their father.